CRANIAL NERVES

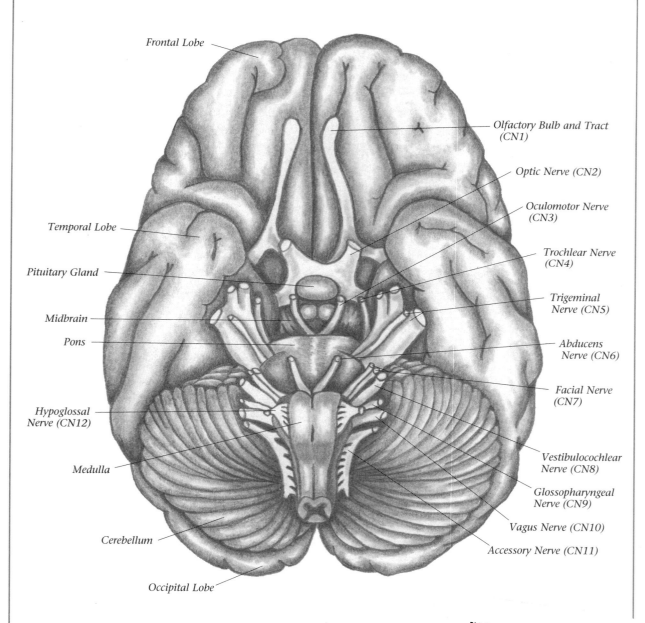

Frontal Lobe

Olfactory Bulb and Tract (CN1)

Optic Nerve (CN2)

Oculomotor Nerve (CN3)

Temporal Lobe

Trochlear Nerve (CN4)

Pituitary Gland

Trigeminal Nerve (CN5)

Midbrain

Abducens Nerve (CN6)

Pons

Facial Nerve (CN7)

Hypoglossal Nerve (CN12)

Vestibulocochlear Nerve (CN8)

Medulla

Glossopharyngeal Nerve (CN9)

Vagus Nerve (CN10)

Cerebellum

Accessory Nerve (CN11)

Occipital Lobe

Screening
Adult Neurologic
Populations

Screening Adult Neurologic Populations

A STEP-BY-STEP INSTRUCTION MANUAL

Sharon A. Gutman, PhD, OTR/L
Alison B. Schonfeld, OTR/L

AOTA PRESS

The American
Occupational Therapy
Association, Inc.

Mission Statement
The American Occupational Therapy Association advances the quality, availability, use, and support of occupational therapy through standard-setting, advocacy, education, and research on behalf of its members and the public.

AOTA Staff
Joseph C. Isaacs, CAE, Executive Director
Karen C. Carey, CAE, Associate Executive Director, Membership, Marketing, and Communications
Jeffrey Finn, Director of Communications

Chris Davis, Managing Editor, AOTA Press
Suzanne Seitz, Production Editor
Barbara Dickson, Editorial Assistant

Robert A. Sacheli, Manager, Creative Services
Sarah E. Ely, Book Production Coordinator

American Occupational Therapy Association, Inc.
4720 Montgomery Lane
PO Box 31220
Bethesda, MD 20824-1220
Phone: 301-652-AOTA (2682)
TDD: 800-377-8555
Fax: 301-652-7711
www.aota.org
To order: 1-877-404-AOTA (2682)

Disclaimers
This publication is designed to provide accurate and authoritative information in regard to the subject matter covered. It is sold or distributed with the understanding that the publisher is not engaged in rendering legal, accounting, or other professional service. If legal advice or other expert assistance is required, the services of a competent professional person should be sought.
—From the Declaration of Principles jointly adopted by the American Bar Association and a
  Committee of Publishers and Associations

It is the objective of the American Occupational Therapy Association to be a forum for free expression and interchange of ideas. The opinions expressed by the contributors to this work are their own and not necessarily those of either the editors or the American Occupational Therapy Association.

ISBN: 1-56900-140-5

Design by Sarah E. Ely
Composition by The Clarinda Company, Atlantic, Iowa
Printed by Victor Graphics, Baltimore, Maryland

Contents

Foreword

ANN BURKHARDT, MA, OTR/L, FAOTA, BCN

For many allied health care professionals, there has been great delight in celebrating the art of clinical practice. Since the emergence of managed care within the American health care system over the past decade, we have found ourselves trying to provide some degree of quality of care, often within a much narrower scope than desirable. The ethics of health care delivery have been repeatedly tested under managed care, and today's marketplace has begun to emphasize the science of practice more than its art. Health care providers currently find themselves accountable for providing evidence-based practice, yet the average clinician has not engaged routinely in clinical research. A clinician's knowledge of "what works in treatment and why" is usually based on anecdotal information from cases, often passed through oral history from therapist to therapist.

As a professional group, occupational therapists have prided ourselves on our ability to capture the subtleties in each case through the narrative process. We have celebrated our ability to tell a client's story based on qualitative evidence of a kind. Mattingly and Fleming (1993) described the clinical reasoning process for occupational therapists and categorized several steps: procedural reasoning, conditional reasoning, and narrative reasoning. Before any story can be told "in good faith," it must be based on evidence. Through the development of procedural reasoning, occupational therapists gain insight into the possibilities for a client based on facts that frame the clinical picture of care.

The formation of procedural reasoning relies not only on facts contained in a history but also on facts the therapist gathers through the screening and evaluative process. Occupational therapists have historically used scales borrowed from other professional groups, because our profession does not have a broad enough foundation in standardized testing. We must rely on some of these scales to assist us to reason clinically and then share the results with other disciplines.

This book is rich in content about standardized tests that are used in interdisciplinary settings, as well as tests that are specific to occupational therapy. Furthermore, this resource provides a consistent, logical, and bulleted decision-making format that easily leads beginning to intermediate-level practitioners into the clinical reasoning and rationalization process. Each section identifies a concept to be assessed, followed by an outline of the functional problems a person could experience. The severity of impairment level is first graded, and then standardized tests and measures are suggested that can be used to perform an accurate assessment of ability. The text also discusses the differences in clinical observation between high-level and low-level functional ability. It provides red flags that assist therapists to rationalize the level of impairment and its relationship to the ability of the person to function safely alone or at home.

One of the most useful items in the book is the resource lists of standardized assessments at the end of each section, including the population for which they are standardly used and the sources from which they can be ordered and obtained. This level of information is often elusive to the average clinician in practice. Many postings on Internet occupational therapy listservs are from clinicians seeking to identify standardized screening tools, where they can be ordered, and the groups or populations to which they are applicable as well as the clinicians' experience with the tools. This feature saves time and centralizes information for clinical managers, advanced clinical staff, and educators who seek to develop or revamp resource tools for teaching therapy or medical students about functional measurement tools.

The cognitive section provides clear, conceptual definitions that are useful for clinicians in identifying a specific functional cognitive impairment that may be affecting their clients' ability to be safe in their personal and physical environments. Emphasis is placed on clients who are functioning at a higher level and the subtleties that can be observed. Many of the resources for health care providers focus on clients functioning at a low level yet, in the experience of average clinicians in general practice, most of the client referrals for cognitive impairment are subtle but are crucial concerning safety. For example, the older person who has surgery and then becomes forgetful and, who also has a decreased attention span and slowed response time will need clearance to drive on hospital discharge, even though it is not safe for the person to do so. However, the person may still perform well on the low-level-functioning screens. The chart-ready forms at the end of the chapter provide a means for clinicians to summarize overall cognitive functional ability and are great for educators seeking to teach students how to do a summary review of the overall cognitive assessment.

Another strength of the cognitive section is the suggestions for assessing the client's ability to respond in emergency circumstances. Many inpatient rehabilitation screens require this type of information in assessment, but all too often it is solely assessed on the basis of asking the client what to do in the event of an emergency, with the therapist satisfied to hear "call 911." That is not enough.

The visual section is helpful not only in its clear text explanation but also through the use of photographs and illustrations demonstrating assessment methods. Instructions are included with pictures demonstrating the concept, for example, of strabismus, visual tracking, and saccadic eye movement screening. Assessment tools are specific to vision and also refer readers to tools specific to occupational therapy.

The section on perception concretizes information that many beginning and experienced general practitioners often find confusing and worrisome in their level of confidence in assessment and the clinical-reasoning process. Perceptual disorders are clearly categorized according to the underlying neurological systems that can be involved in the overall functional presentation of the impairment. Examples of axes of dysfunction are provided, such as visual perception, visual–spatial perception, tactile perception, body schema perception, language perception, and motor perception disorders. These are demonstrated functionally, based on what clinicians observe during a top-down clinical assessment while engaging in a functional task. The section includes assessments that are commonly used in medical settings by traditional practice partners (e.g., medical doctors, neuropsychologists, physical therapists, and speech pathologists) that also can be used to tease out the underlying functional diagnosis. This melding of the traditional with the contextual is helpful in development of clinical reasoning with perceptual disorders in general.

The motor section contains information on classical, standardized, and cross-disciplinary means by which rehabilitation professionals usually measure range of motion, strength, and reflex. This section takes readers a step beyond the traditional, because it also pairs the standard content with occupational therapy content that is specific to function. One useful aspect of the functional tables is the correlation of the numeric range of motion and strength grade estimates and functional tasks. This streamlines the reporting process for clinicians relative to the meaning that diminished motor functioning can have on a person's functional activities and participation in daily life tasks. As a manager, I can visualize the use of these charts to help problem solve templates for electronic documentation. As an educator, I can foresee their usefulness in helping beginners develop clinical reasoning and make the cognitive-learning connection between strength and task application. As a researcher, I notice that access to all of the scales specific to occupational therapy is available, as well as the well-researched cross-discipli-nary scales of activities of daily living functioning. Being more research-ready will allow occupational therapists to gain competence as entry-level research clinicians, whose views are included in contributions to evidence that can be later used to shape health policy and reimbursement decisions.

The book distinguishes the functions of the cerebellum and basal ganglion from the general motor assessment. It is helpful to separate motor dysfunctions of the higher cerebral cortex from those of the lower, and evolutionarily older, regions of the central nervous system. The section clearly explains the impact on fluidity of movement and precision of movement when dysfunction occurs. The explanation of dystonia and the glossary of torsion dystonia terms are useful taxonomic tools. Occupational and physical therapists are often asked to evaluate and treat dystonia patients, and only a few resources exist on this particular diagnostic category. In the red flags section is the often-overlooked sign of motor impersistence; this sign can help when assessing patients who extinguish attention to their limbs even when engaged in functional tasks, such as holding a sandwich or a feeding or grooming utensil.

Cerebella and basal ganglia signs are also important for their contribution to the risk of falling, which is a major indicator of morbidity and mortality in all people older than age 65, particularly those who have a neurological disease or diagnosis, and is an issue in facilities and community-based care for older people. The in-depth assessment section contains several valuable references for standardized tests that are specific to measuring disorders of balance as well.

The cranial nerve function section is succinct and straightforward. Direct correlation is, once again, drawn to functional impairments associated with neurological dysfunction. The photographs are helpful in demonstrating clinical tests of cranial nerve signs. The chart-friendly screening tool is the best published to date, containing check-off boxes and some direct corre-lation to function, which is a helpful prototype for a manager or a supervisory clinician to use to perhaps develop templates for electronic documentation systems.

Peripheral neuropathy is one of the least represented resources throughout most rehabilitation textbooks and yet one of the most frequently encountered neurological problems. Some people who have diabetes have peripheral neuropathy; also people who have cancer and are receiving chemotherapy treatment and people who have immunosuppressive disorders often have peripheral neuropathy. Older people who have back problems with spinal nerve compression, such as spinal stenosis, also experience it. The side effects of peripheral neuropathy include intractable burning pain; thus, the presence of neuropathy moderately to severely limits its activity participation.

The sensory impairment indications list is clear and brief. Brachial plexopathy, one clinical sign that is often overlooked or misdiagnosed, is included. The photographs show the nuances of brachial plexopathy assessment, including reflex measurement. The explanation of the difference between reflex sympathetic pain and peripheral nerve pain is understandable. The soft tissue integrity section is also helpful, because skin changes are common.

The dysphagia chapter provides a stepwise clinical reasoning approach to assessment of dysphagia, or swallowing disorders. There are comprehensible descriptors of the differences between lingual (tongue strength or sensory) impairment versus mandibular (jaw strength impairment) as well as clear delineation between the oral and pharyngeal stages of the swallowing reflex. There is also a good differentiation between the most commonly used standardized forms of radiological testing: the modified barium swallow and flexible endoscopy.

Dysphagia is an area of practice that is crucial to prognosis and length-of-stay issues in facilities. Populations that can experience aspiration pneumonia as a side effect of swallowing impairment include the adults who have had a stroke, people with head injuries, and people with progressive neurological disorders.

To conclude, this text is an extremely helpful, concise reference that most clinicians, educators, managers, and students will find time in clinical and educational settings. The book succinctly compiles and centrally locates a plethora of information and resources for practitioners who assess the presence and functional impact of neurological diseases and conditions in the medical model setting or community. Also, because of its format—lists, tables, photographs, and bulleted sections—this text will be useful in academic settings to educators as well as to students of neurorehabilitation. As a neurorehabilitation specialist, I look forward to having a copy in my own professional library and will recommend it to our campus library and bookstore.

Reference

Mattingly, C., & Fleming, M. H. (1993). *Clinical reasoning: Forms of inquiry in a therapeutic practice.* Philadelphia: F. A. Davis.

Introduction

This book aims to be a functional and easy-to-understand instruction manual of the most common screening methods for neurologic patient populations. It is intended to allow therapists to easily identify possible dysfunction, document identified dysfunction, and determine if further in-depth evaluation is warranted. Students will find this book to be a valuable reference tool to take from the classroom to the clinic. Therapists will also find this book to be a unique reference tool frequently used in the clinic and home care settings.

In the process of assessment there are two primary steps: (a) screening and (b) evaluation. A screening is a quick, cost-efficient method of detecting the presence of a deficit area. If a screening method detects a deficit area, an in-depth evaluation tool is then administered to better discern (a) the degree of severity of the deficit and (b) how the deficit has impaired the patient's daily life function.

A screening method is not used for diagnostic purposes or treatment planning. Only the results of an in-depth evaluation can be used to make a diagnosis and formulate treatment plans. Rather, a screening method allows a therapist to detect the presence of a deficit quickly. Screening methods also facilitate a therapist's ability to select the most appropriate in-depth evaluation tools that can best assess specific deficit areas—initially detected by the screen—in greater detail.

Because the administration of in-depth evaluation tools is time and labor intensive, therapists must be able to select evaluation tools that can precisely target a patient's specific deficit areas. The ability to select the best in-depth evaluation tool is dependent on the appropriate use of screening tools. In this era of managed health care, the appropriate use of quick and cost-efficient screening methods has become critical in the assessment process, as the time allotted for the assessment process has been significantly reduced. Therapists must be able to quickly pinpoint deficit areas (using screening methods) and evaluate how those deficit areas have impaired the patient's daily life function (using in-depth evaluation tools). Sometimes a screening detects the possible presence of a pathological condition that warrants further in-depth evaluation by a physician or other health care professional. When this occurs, a referral is made to the appropriate health care professional.

It should be noted that the use of screening methods is limited to registered or licensed occupational therapists. Certified occupational therapy assistants are not licensed to administer or interpret the results of screenings.

This book describes methods of screening for an adult neurologically involved population (e.g., those with traumatic brain injury, spinal cord injury, cerebrovascular accident, multiple sclerosis, amyotrophic lateral sclerosis, dementia, Parkinson's disease). Screening methods for a pediatric neurologic population are not covered.

Nine primary areas of neurologic screening are presented:

- Cognition
- Vision
- Perception
- Sensation
- Peripheral nerve function
- Motor function (including deep tendon reflex function)
- Basal ganglia and cerebellar function (balance, postural control, automated movements)
- Cranial nerve function
- Dysphagia.

Each of the nine screening sections are divided into four subsections:

1. "Functional Implications of Impairment"
2. "Screening Procedures"
3. "Red Flags: Signs and Symptoms"
4. "Available In-Depth Assessments."

In the section, "Functional Implications of Impairment," readers can find information about how patients are affected by a specific neuropathology with regard to their everyday functioning in daily life activities. This section will help therapists and students recognize the specific types of difficulties patients will have in functional daily life activities as a result of neuropathology.

The section "Screening Procedures" describes the step-by-step instructions required to carryout each screen. The steps are written in an easy-to-follow format that can be used right in the clinic. Photographs are used to illustrate specific screening procedures. A description of patient responses indicating impairment accompany each set of screening procedures so that therapists and students can differentiate possible dysfunction from function. Patient responses can be recorded right on the screening form that accompanies each screen (at the end of each section).

"Red Flags: Signs and Symptoms," describes the common signs and symptoms that patients present as a result of specific neuropathology. Therapists and students can use this section to easily identify presenting symptoms of neurologic conditions.

"Available In-Depth Assessments" is a section that exhaustively lists, in alphabetical order, possible in-depth evaluations to use if a screen detected possible impairment and if use of an in-depth evaluation is warranted. Therapists and students can also use this section to find the source of a specific evaluation in order to obtain it.

At the end of each section, a screening form can be found on which to record patient responses for each specific screen. The screening form corresponds to the actual steps outlined in the "Screening Procedures" section, so that a therapist can administer a screen using the step-by-step instructions and record patient responses on the screening form (screening forms are meant to be copied for multiple use in the clinic).

As nearly one third of all patients have neurologic disorders, this book provides a needed resource for therapists and students who work with neurologic patients and who need to learn the step-by-step screening methods to identify possible dysfunction. It is a valuable resource that can be used in the classroom and clinic throughout a therapist's career.

Acknowledgments

We thank our colleagues Kathryn Robshaw, PT; Eric Schwabe, PT; Richard Sabel, MS, OTR/L; Marianne Mortera, MS, OTR/L; Glen Gillen, MPA, OTR/L, BNC; and Leslie Kane, MS, OTR/L. We also thank Mount Sinai Hospital of New York City.

DEDICATION

This book is lovingly dedicated to my parents Sy and Deena and my friends who supported and encouraged me that "I can do it."

—Alison B. Schonfeld, OTR/L

Cognitive
SCREENING

FUNCTIONAL IMPLICATIONS OF COGNITIVE IMPAIRMENT

Cognition is the ability to use mental processing skills to interact with and meet the demands of one's environment. It is mediated by the *frontal lobes* of the brain. Mental processing skills extend on a continuum from low-level skills (e.g., alertness/arousal, orientation, recognition, initiation/termination of activity) to more complex, higher-level skills (e.g., sequencing multiple steps, categorizing multiple items, problem solving, planning, generalization of new learning, insight).

Patients whose low-level cognitive skills are impaired will likely present the following:

- *Low level of alertness/arousal.* The patient may be in a stuporous state and be unaware of others in his or her environment.

- *Disorientation to person, place, and/or date.* The patient may be unable to state accurately what yearly season it is.

- *Lack of ability to recognize familiar others in one's environment.* The patient may be unable to recognize loved ones.

- *Inability to concentrate on a basic activities of daily living (ADL) task for a length of time greater than 1 to 2 minutes.* The patient may be unable to complete his or her morning grooming routine due to distractibility.

- *Inability to initiate ADL.* The patient may be unable to get out of bed in the morning without cues; he or she may be unable to initiate dressing and may sit in his or her chair for hours until further help is offered.

- *Inability to sequence the steps of a basic ADL task accurately.* The patient may inaccurately sequence the steps of brushing teeth; for example, he or she may put toothpaste on the brush before water or may brush only the upper teeth and forget to brush the lower teeth.

- *Inability to categorize items accurately.* The patient may be unable to sort items such as forks, spoons, and knives.

- *Inability to follow one- or two-step commands or directions accurately.* The patient may be unable to follow simple directions such as "Pick up the hairbrush and brush your hair."

- *Poor safety/judgment.* The patient may attempt to get out of bed or rise from a wheelchair despite having poor balance, ataxia, and incoordinated gross motor movements that increase his or her risk of falls and further injury.

Patients whose high-level cognitive skills are impaired may present the following:

- *Inability to sequence the steps of a basic ADL task accurately.* The patient may inaccurately sequence the steps of dressing; for example, he or she may put shoes on before socks or don pants before underwear.

- *Inability to categorize items accurately.* The patient may be unable to sort laundry items into separate loads for colors and whites and, thus, ruin many clothing items.

- *Inability to follow two- or three-step commands or directions accurately.* The patient may be unable to follow the directions to cook a microwave dinner.

- *Short-term memory deficits.* The patient may be unable to remember what he or she needed at the store, the time of day or night to take medication, or the names of people just met at a new job.

- *Difficulty problem solving.* The patient may be unable to problem solve how to cross the street at a four-way intersection, or he or she may be unable to problem solve how to balance a checkbook, pay bills on time, and budget a monthly social security income check.

- *An inability to demonstrate generalization of learning.* The patient may have learned to use the microwave accurately at home but be unable to generalize those skills to the microwave at work that has a slightly different arrangement of control buttons.

- *Lack of insight.* The patient may display a lack of insight regarding social skill deficits and repeatedly use inappropriate greetings in a public place.

- *Poor safety/judgment.* The patient may forget to use oven mitts when removing food from the oven and burn his or her hands, or a patient may forget that it is unsafe to smoke in bed and fall asleep with a lit cigarette, thus starting a fire.

COGNITIVE SCREENING PROCEDURES

A basic screening for *mental status* should always be administered *first* in the evaluation process because a patient's cognitive status will affect his or her subsequent performance on all other screening procedures. Sometimes it is difficult to distinguish between a *cognitive impairment* and a *perceptual impairment.* For example, a patient who demonstrates difficulty sequencing the steps of dressing may possess either an impairment in cognition or a perceptual–motor planning deficit. In such cases, more in-depth evaluation is required. It also is important that the therapist choose screening items and occupations that are *culturally familiar* to the patient; otherwise, the therapist will have difficulty distinguishing between a patient's cognitive deficits and lack of familiarity with screening activities and objects.

Screening for cognitive deficits should be separated by high- and low-level cognitive skills. Cognitive screenings allow therapists to determine whether a deficit area exists and whether an in-depth evaluation should be administered. A cognitive screening will reveal whether the patient's mental processing skills are normal or whether a possible cognitive impairment is detected.

A cognitive screening for a patient with low-level cognitive skills may focus on the following:

- Arousal

- Attention

- Orientation

- Recognition

- Simple command following

- Memory

- Initiation of activity.

In addition, a cognitive screening for a patient with high-level cognitive skills may focus on higher executive functions, including the following:

- Insight

- Multiple-step command following

- Mental flexibility

- Planning

- Problem solving

- Abstraction

- New learning

- Generalization of learning

- Safety/judgment.

GLASGOW COMA SCALE

The Glasgow Coma Scale is commonly used to measure levels of consciousness following a traumatic injury to the brain. The 15-point scale relates levels of consciousness to eye opening, motor response, and verbal response. Patients scoring 8 or less are considered to have sustained severe brain damage. Scores ranging from 9 to 12 indicate moderate brain damage, and scores from 13 to 15 indicate mild brain damage.

Examiner's Test	Patient's Response	Score
Eye opening		
Spontaneous	Opens eyes on own	4
To speech	Opens eyes when asked to in a loud voice	3
To pain	Opens eyes when pinched	2
	Does not open eyes when pinched	1
Best motor response		
To commands	Follows simple commands	6
To pain	Pulls examiner's hand away when pinched	5
	Pulls a part of body away when pinched	4
	Flexes body part inappropriately to pain (decorticate posturing)	3
	Body becomes rigid in an extended position when examiner pinches patient (decerebrate posturing)	2
	No motor response to pinch	1
Verbal response		
To speech	Converses appropriately with examiner; oriented to person, place, time	5
	Appears confused or disoriented	4
	Speaks clearly but makes no sense	3
	Makes incomprehensible sounds	2
	Makes no sound	1

Note. From *Introduction to Clinical Neurology, 2nd Edition* (p. 273), by D. J. Gelb, 2000, St Louis, MO: Elsevier Science. Copyright 2000 by Elsevier Science. Adapted with permission.

RANCHO LEVELS OF COGNITIVE FUNCTIONING–REVISED

The Rancho Levels of Cognitive Functioning–Revised is a useful scale for describing a patient's general cognitive and behavioral status (after head injury).[1]

I. No Response, *Total Assistance*
Patient appears to be in a deep sleep and is completely unresponsive to any stimuli.

II. Generalized Response, *Total Assistance*
Patient reacts inconsistently and nonpurposefully to stimuli in a nonspecific manner. Responses are limited and often the same regardless of stimuli presented. Responses may be physiologic changes, gross body movement, and vocalizations. Often, the earliest response is to deep pain. Responses are likely to be delayed.

---

[1]This scale is authored by Chris Hagen, PhD, CCC-SLP (as cited in Neistadt, 2000).

III. Localized Response, *Total Assistance*

Patient reacts specifically but inconsistently to stimuli. Responses are directly related to the type of stimulus presented, as in turning the head toward a sound or focusing on an object presented. The patient may withdraw an extremity and/or vocalize when presented with a painful stimulus. He or she may follow simple commands in an inconsistent, delayed manner, such as closing eyes or squeezing or extending an extremity. After external stimulus is removed, the patient may lie quietly. He or she also may show a vague awareness of self and body by responding to discomfort by pulling at nasogastric tube or catheter or resisting restraints. The patient may show bias by responding to some persons (especially family and friends) but not to others.

IV. Confused–Agitated, *Maximal Assistance*

Patient is in a heightened state of activity with severely decreased ability to process information. He or she is detached from the present and responds primarily to his or her own internal confusion. Behavior is frequently bizarre and nonpurposeful relative to his immediate environment. The patient may cry out or scream out of proportion to stimuli even after removal, show aggressive behavior, or attempt to remove restraints or tubes or crawl out of bed in a purposeful manner. He or she does not, however, discriminate among persons or objects and is unable to cooperate directly with treatment efforts. Verbalizations frequently are incoherent and/or inappropriate to the environment. Confabulation may be present; he or she may be euphoric or hostile. Thus, gross attention is very short, and selective attention is often nonexistent. Being unaware of present events, patient lacks short-term recall and may be reacting to past events. He or she is unable to perform self-care (feeding and dressing) without maximum assistance. If not physically disabled, he or she may perform motor activities as in sitting, reaching, and ambulating but as part of his or her agitated state and not necessarily as a purposeful act or on request.

V. Confused–Inappropriate Nonagitated, *Maximal Assistance*

Patient appears alert and is able to respond to simple commands fairly consistently. However, with increased complexity of commands or lack of any external structure, responses are nonpurposeful, random, or fragmented toward any desired goal. He or she may show agitated behavior, not on an internal basis (as in Level IV) but as a result of external stimuli, and usually out of proportion to the stimulus. The patient has gross attention to the environment but is highly distractible and lacks ability to focus attention to a specific task without frequent redirection back to it. With structure, he or she may be able to converse on a social, automatic level for short periods of time. Verbalization is often inappropriate; confabulation may be triggered by present events. Memory is severely impaired, with confusion of past and present in reaction to ongoing activity. The patient lacks initiation of functional tasks and often shows inappropriate use of objects without external direction. He or she may be able to perform previously learned tasks when structured for him or her but is unable to learn new information. Patient responds best to self, body, comfort, and often family members. The patient can usually perform self-care activities with assistance and may accomplish feeding with maximum supervision. Management on the ward often is a problem if the patient is physically mobile, as he or she may wander off either randomly or with vague intention of "going home."

VI. Confused–Appropriate, *Moderate Assistance*
Patient shows goal-directed behavior but is dependent on external input for direction. Response to discomfort is appropriate, and he or she is able to tolerate unpleasant stimuli (e.g., NG [nasogastric] tube) when need is explained. Patient follows simple directions consistently and shows carryover for tasks he or she has relearned (e.g., self-care). He or she is at least supervised with old learning and unable to maximally assist for new learning with little or no carryover. Responses may be incorrect due to memory problems, but they are appropriate to the situation. They may be delayed, and patient shows decreased ability to process information with little or no anticipation or prediction of events. Past memories show more depth and detail than recent memory. The patient may show beginning awareness of his or her situation by realizing he or she does not know an answer. He or she no longer wanders and is inconsistently oriented to time and place. Selective attention to tasks may be impaired, especially with difficult tasks and in unstructured settings, but is now functional for common daily activities (e.g., 30 minutes with structure). Patient shows at least vague recognition of some staff members and has increased awareness of self, family, and basic needs (as food), again in an appropriate manner as in contrast to Level V.

VII. Automatic–Appropriate, *Minimal Assistance for Routine Daily Living Skills*
Patient appears appropriate and oriented within hospital and home settings, goes through daily routine automatically (but frequently robot-like) with minimal to absent confusion, but has shallow recall of what he or she has been doing. He or she shows increased awareness of self, body, family, foods, people, and interaction in the environment. The patient has superficial awareness of but lacks insight into his or her condition, demonstrates decreased judgment and problem solving, and lacks realistic planning for the future. He or she shows carryover for new learning, but at a decreased rate. The patient requires at least minimal supervision for learning and for safety purposes. He or she is independent in self-care activities and supervised in home and community skills for safety. With structure, the patient is able to initiate tasks in social and recreational activities in which he or she now has interest. Judgment remains impaired, such that he or she is unable to drive a car. Prevocational or avocational evaluation and counseling may be indicated.

VIII. Purposeful and Appropriate, *Standby Assistance*
Patient is alert and oriented, is able to recall and integrate past and recent events, and is aware of and responsive to his or her culture. He or she shows carryover for new learning if acceptable to him or her and his or her life role and needs no supervision after activities are learned. Within physical capabilities, patient is independent in home and community skills, including driving. Vocational rehabilitation, to determine ability to return as a contributor to society (perhaps in a new capacity), is indicated. He or she may continue to show decreased abilities relative to premorbid abilities, reasoning, tolerance for stress, judgment in emergencies, or unusual circumstances. Social, emotional, and intellectual capacities may continue to be at a decreased level but are functional for society.

IX. Purposeful and Appropriate, *Standby Assistance on Request*
Patient can independently shift between learned tasks and can complete tasks without error for approximately 2 hours. He or she requires assistive memory devices to recall

daily schedule, such as a to-do list and a daily planner. When requested, he or she can record important information for later use with minimal assistance. He or she can independently carry out familiar self-care, work, and leisure activities but requires minimal assistance with new, unfamiliar ADL. Patient is aware of and acknowledges impairments when they interfere with ADL—he or she can also appropriately correct errors once recognized. However, patient requires standby assistance to anticipate a problem before it occurs in order to act to prevent it. When requested, he or she also is able to think about the consequences of his or her actions and behaviors with standby assistance. Patient accurately estimates abilities and limitations but requires minimal assistance to adjust to new task demands. With minimal assistance he or she is able to acknowledge others' needs and feelings and respond appropriately. He or she also is able to self-monitor the appropriateness of his or her social interactions with standby assistance. Low frustration tolerance may be problematic; the patient may become easily agitated. Depression may continue.

X. Purposeful and Appropriate, *Modified Independent*
Patient is able to carry out multiple tasks simultaneously in all environments but likely requires periodic breaks. He or she is able to procure and maintain his or her own assistive memory devices. ADL activities are carried out independently, but patient may require more than the usual amount of time or compensatory strategies. The patient can anticipate problems before they occur and take action to prevent them but may require more than the usual amount of time or compensatory strategies. He or she is able to think about the consequences of his or her behaviors but may require greater than usual time or compensatory strategies to select an appropriate choice of action. The patient can estimate his or her abilities and limitations accurately and can adjust independently to new task demands. He or she also is able to recognize the needs and feelings of others and can independently respond in an appropriate manner. Social interaction skills are consistently appropriate; however, the patient may experience irritability and low frustration tolerance when sick, fatigued, or under emotional stress. Periodic episodes of depression may continue.

COMATOSE/STUPOROUS PATIENT

Coma is a state in which the patient experiences a loss of consciousness characterized by a loss of awareness of self and environment and an inability to respond to external stimuli or internal drives. There are differing degrees of coma. All of the following states of coma are characterized by both a reduced receptivity to stimulation and a reduced responsivity.

- *Profound coma* is characterized by complete unresponsiveness to the environment. Even the most painful stimuli have no effect on the patient.

- *Semi-coma* is a lighter state of coma characterized by groaning, stirring, increased respiration, or a brief opening of the eyes in response to painful stimuli (pinch).

- *Stupor* is a still lighter state of coma in which the patient will open his or her eyes and make some simple response to stimuli (e.g., loud voice, pinch), but the patient does not speak.

■ *Minimally conscious or a drowsy–confused coma* state is one in which the patient will speak but is unable to respond appropriately or process thoughts with normal speed and clarity. A patient who is drowsy–confused has a tendency to fall into an inattentive, stupefied state if left unstimulated.

When a comatose patient has full receptivity to external stimuli but lacks the ability to make a response, the condition is known as *locked-in syndrome*. This condition most often is due to a lesion in the basil pons that interrupts the descending motor pathways but spares the ascending sensory pathways. If the patient lacks the impulse to move, although not paralyzed, the condition is referred to as *catatonia* or *abulia*.

In a *persistent vegetative state*, the patient is in a state of eyes-open unconsciousness. The patient is able to blink to threat and is capable of a few primitive postural movements but is otherwise without awareness and unresponsive to external stimuli. In persistent vegetative states, the brain stem and its vegetative functions (i.e., cough, gag, and swallowing reflexes) are spared. The brunt of neurologic damage is to the cerebral hemispheres. Thus, life expectancy is lengthened—the patient can remain alive in this state indefinitely with technological life support.

Brain death occurs when all brain stem functions are lost. This type of coma is a state of sleep-like (eyes-closed) unarousability due to extensive damage to the reticular activating system. All vegetative functions are lost, leading to fatal respiratory infections.

When screening the drowsy–confused comatose patient for cognitive function, the following procedures are used:

1. *Level of arousal*. Do the patient's eyes open spontaneously when his or her name is called in a loud voice or when physically pinched?

2. *Orientation*. Does the patient remember his or her name? Does the patient know that he or she is in the hospital? Does he or she know what season of the year it is?

3. *Recognition*. Is the patient able to recognize family members and friends? Is he or she familiar with hospital staff members?

4. *Amnesia*. Can the patient recall memories from his or her past, for example, How old are you? Where did you go to school? Where do you live? *Retrograde amnesia* is the loss of one's entire personal past after traumatic injury to the brain. It is a common consequence of brain injury. Long-term memory often returns to patients as they recover from brain injury. *Anterograde amnesia* is a decreased ability to remember ongoing day-to-day experiences after the injury has occurred. It involves a decreased ability to encode short-term memory into long-term storage.

5. *Attention*. Is the patient able to attend to the voice of the therapist? Is the patient able to attend to simple tasks (e.g., visually tracking a brightly colored object, washing face with a washcloth)? Or, does the patient become easily distracted by other stimuli (e.g., voices in the hallway)?

6. *Command following*. Is the patient able to follow simple one-step commands, for example, Try to keep your eye on the moving ball. Try to reach for the red block.

PATIENT WITH LOW-LEVEL COGNITIVE FUNCTIONING

When screening the patient with low-level cognitive functioning, the following procedures are used.

1. *Alertness/arousal.* Is the patient alert and aware of his or her surroundings?

2. *Orientation.* Is the patient oriented to self, place, and time? Ask the patient the following:

   - What is your name? Where do you live?

   - Do you know where you are now [hospital]? Do you remember why you are in the hospital?

   - Do you know what time of the day it is? Do you know what the date is? (Often, patients in the hospital will lose track of the day of the week; it is more significant if the patient cannot accurately identify the month, season, or year.) Do you know what month this is? Season? Year?

☞ Cognitive impairment is indicated if the patient is unable to answer more than half of these questions.

3. *Recognition.*

   - Present the patient with photos of family members, famous people, or people featured in the current day's newspaper.

   - Ask the patient to identify their names. If the patient cannot identify their names, give him or her a choice of three options.

   - Ask the patient whether he or she can remember his or her relationship to the photographed family members, or ask the patient what the famous photographed individuals are known for.

☞ Cognitive impairment is indicated if the patient is unable to recognize close family members or if he or she is unable to recognize more than half of the photos of familiar famous people.

4. *Attention/concentration.* Note whether the patient is able to consistently attend to the screening procedures and whether he or she becomes distracted by extraneous noise in the environment (e.g., the conversations of other people). *Selective attention* is the ability to attend to one task while blocking out distracting events in the environment. For example, can the patient attend to the therapist's instructions while blocking out the conversation of nearby others? A common phenomenon of brain injury is the incorporation of words heard in someone else's conversation. The patient is unable to selectively attend to his or her own conversation with the therapist and begins to incorporate words heard from the conversation of others in the environment.

☞ Cognitive impairment is indicated if the patient consistently becomes distracted and requires redirection.

5. *Initiation of activity.* Is the patient able to initiate ADL independently, or does the patient require cues and setup? Lack of initiation of activity often results from dorsolateral frontal lobe damage. Such patients often present with flat affect and display little motivation to participate in once-desired activities. If the patient is able to initiate an ADL with cues and setup, is the patient able to complete the activity independently, or does he or she require continued cues? For example, a patient who is given cues to initiate dressing may dress as long as the therapist is present to provide cues. Left unattended, the patient may remain seated in his or her chair but fail to continue the dressing activity.

- Instruct the patient to initiate a simple ADL activity.

- Present the patient with a loaf of bread, butter, and a knife (or blunted spreader if the patient appears aggressive).

- State to the patient, "Show me what you would do with these items."

- Note whether the patient is able to initiate the activity without further cues.

☞ Cognitive impairment is indicated if the patient is unable to butter one piece of bread without further cues.

6. *Termination of activity.* Once started on an ADL, can the patient terminate the activity when completed, or does the patient continue to carry out the activity without external cues to stop? *Perseveration* occurs when a patient is unable to terminate participation in an activity. For example, during a morning grooming activity, a patient who is set up to brush her teeth may fail to stop brushing her teeth without cues to terminate the activity. Note whether the patient was able to terminate the above activity (i.e., buttering a piece of bread) independently.

☞ Cognitive impairment is indicated if the patient is unable to stop buttering slices of bread without verbal cues.

7. *Long-term memory (LTM) and short-term memory (STM).* LTM, or remote memory, is often spared or returns as the patient recovers from brain injury. LTM is believed to be mediated by limbic system structures deep in the core of the brain—most notably the hippocampus. Ask the patient at least four of the following questions:

- What are the names of your children and spouse?

- In what year were you born?

- In what year were you married?

- What was your mother's maiden name?

- Where did you grow up?

- What year did you graduate from school?

☞ Cognitive impairment is indicated if the patient cannot answer more than two of the four questions.

If STM, or recent memory, is impaired or lost after brain injury, it usually remains impaired. STM is believed to be mediated by frontal lobe structures and is directly related to attention. Ask the patient questions about events that occurred in the past 24 to 48 hours. For example,

- What did you have for breakfast today?

- Did you have visitors last evening? Who were they?

- What is my name? When did you first meet me?

- When did you arrive at the hospital? How many days have you been in the hospital?

☞ Cognitive impairment is indicated if the patient cannot answer more than half of the above questions.

- Instruct the patient that you are going to ask him or her to remember three items (e.g., a watch, a necktie, a coffee mug).

- State the three items and tell the patient that you are going to ask him or her to recall those items after a short time has passed.

- Go on to test the patient's fund of knowledge.

- After 3 to 5 minutes have passed, ask the patient to recall and name the three items. The ability to reproduce these items after a short interval is a test of *retentive memory span*. Patients who demonstrate good memory can reproduce two or three items.

☞ Cognitive impairment is indicated if the patient reproduces only one (or no) item.

8. *Fund of knowledge.* Ask the patient at least four of the following questions:

- Who is the current president? Who was the first president? What is the name of the president before the current president?

- What is the name of a large city in the northeastern United States?

- How many weeks are in a year? How many days are in a year?

- Name as many oceans or bodies of water as possible.

- Name as many states, countries, and continents as possible.

☞ Cognitive impairment is indicated if the patient is unable to answer more than two of the four questions.

9. *Sequencing.* Ask the patient to perform at least three of the following:

- Count forward and backward (e.g., from 20 to 30).

- State the months of the year in sequence—forward and backward.

- Describe the steps of laundry (or fixing a flat tire, preparing a peanut butter and jelly sandwich).

- Demonstrate the steps of ordering a pizza from a pizza shop using the telephone, phone book, paper, and pencil provided.

☞ Cognitive impairment is indicated if the patient mixes up the sequence in any two of these instructions.

10. *Categorization.* Ask the patient to name as many items as possible in a specific category. For example, ask three of the following questions:

- Name as many farm animals as possible.

- Name as many types of automobile makers as possible.

- Name as many modes of transportation as possible.

- Name as many words as possible that begin with the letter *P.*

☞ Most adults can name 20 to 25 items in a single category within 1 minute. Cognitive impairment is indicated by an inability to generate more than 3 to 5 items in each of the indicated categories.

The ability to cognitively generate items belonging to a specific category is a test of *concept formation* and *thought generation. Poverty of thought* is characterized by difficulty generating items belonging to a specific category. It is also characterized by a paucity of thought—the individual has difficulty generating thoughts in response to any question. For patients with aphasia, provide a simple categorization task. For example, ask the patient to sort a pile of utensils into groups of spoons, forks, and knives.

11. *Command following.* Instruct the patient to carry out a simple two- or three-step task. For example, ask the patient to perform at least three of the following:

- Take a piece of paper and fold it in thirds. Place the folded paper in the envelope.

- Write your name, address, and social security number on this piece of paper.

- Open the packaged food item and put it in the bowl. Mix up the contents of the bowl with this spoon.

- Tap your knees three times with (one or) both hands. Then clap your hands together three times (or clap your hand on your knee).

☞ Cognitive impairment is indicated if the patient is unable to correctly carry out the two or three directions in two of these tasks.

12. *Safety/judgment.* Is the patient a safety risk to himself or herself and others? For example, is the patient unaware that his or her hemiplegic arm is hanging over the wheelchair armrest, in the wheelchair spokes? Is the patient impulsive? Impulsivity often results from orbitofrontal lobe damage. For example, does the patient attempt to get out of the wheelchair without locking the brakes and moving the footplates away? Does the patient reach for hot items on the stove without using pot holders? Does the patient understand simple emergency procedures to use if an accident occurs? Describe at least four of the following emergency situations and ask the patient to determine what he or she would do.

- You are asleep in your room at home. The smoke alarm goes off. What should you do?

- You lose your keys and are locked out of your house. What should you do?

- You are home alone. You fall and are unable to move your leg. You are in great pain. What's the best thing to do?

- You are preparing dinner. High flames begin to come from the pan on the stove. What should you do?

- You are in a large shopping mall and you become separated from the person you came with. You are lost. What should you do?

☞ Safety may be compromised if the patient demonstrates any of the above behaviors or is unable to offer appropriate solutions for any of the indicated emergency situations.

13. *Self-control.* Self-control is the ability to monitor and regulate one's impulses, urges, and compulsions. Self-control is considered to be one of the *executive functions*—the cognitive skills one needs to override socially unacceptable urges in order to behave in socially appropriate ways.

Disinhibition—or loss of the ability to self-regulate socially inappropriate behaviors—often occurs as a result of orbitofrontal lobe damage. The patient may display impulsivity, aggression, irritability, agitation, and/or sexual disinhibition. During the screening procedures, the therapist should note whether the patient presents with any of the following behaviors of disinhibition:

- Does the patient verbalize inappropriate sexually related comments?

- Does the patient act impulsively? Does the patient act on any impulse that enters his or her consciousness?

- Is it apparent that the patient acts on impulse without considering the consequences of his or her behavior?

- Is the patient verbally or physically aggressive?

- Does the patient present irritability or agitation in response to the slightest provocation?

- Is the patient labile? Does he or she have difficulty controlling his or her emotions? Is the patient frequently tearful? Do the patient's moods change suddenly, without correspondence to environmental events?

☞ Impaired executive functions mediating self-control are indicated by demonstration of any of these above behaviors.

PATIENT WITH HIGH-LEVEL COGNITIVE FUNCTIONING

When screening the patient with high-level cognitive functioning, the following procedures are used:

1. *Arousal/alertness.* The patient with high-level neurological involvement should be aware of his or her surroundings even if his or her affect is one of indifference or appears flat. Patients with dorsolateral frontal lobe damage often present with flat affect but are nevertheless alert and aware of individuals and objects in their environment.

2. *Orientation.* The patient should be oriented to self, place, and time. Ask the patient the following:

- What is your name? Where do you live?

- Do you know where you are now (hospital)? Do you remember why you are in the hospital?

- Do you know what time of day it is? (If the patient does not know the exact time, he or she should be able to approximate the time of day accurately.) Do you know what the date is (the day of the week, the month, the year)?

☞ Cognitive impairment is indicated if the patient is unable to accurately answer more than half of these questions.

3. *Recognition.* The patient should be able to recognize family members, friends, and familiar hospital staff members consistently.

- Present the patient with photos of family members and friends, famous people, and people featured in the current newspaper.

■ Ask the patient to identify their names. If the patient cannot identify their names, offer the patient a choice of three options or provide similar verbal cues (e.g., "This man is a famous actor.").

■ Ask the patient to identify familiar hospital staff members (e.g., What is my name? What is the name of your doctor?).

☞ Cognitive impairment is indicated if the patient is unable to recognize family members, friends, and familiar hospital staff members or if the patient is unable to recognize more than half of the photos of familiar famous people.

4. *Attention/concentration.* The patient should be able to display selective attention—the ability to attend to one task while blocking out extraneous noise (or other stimuli) in the environment. Observe the patient's ability to attend to a task without becoming distracted. For example, ask the patient to perform at least two of the following tasks:

■ Read a billing statement and write a check for the amount due.

■ Read a bus or train schedule and determine which bus or train to take in order to arrive in a specific city at a specific time.

■ Read a map and determine which route to take to travel to a specific town.

Is the patient able to consistently attend to the task without distraction? Does the patient require cues or verbal redirection to return to the task when distracted? How long is the patient able to attend to the task before becoming distracted? How often does the patient require cues or redirection to return to the task when distracted?

☞ Cognitive impairment is indicated if the patient is unable to consistently attend to the task without distraction and requires verbal cues to return to the task.

5. *Mental flexibility.* This is a component of selective attention. It requires the patient to switch back and forth between two (or more) separate tasks. The patient must demonstrate selective attention for each task while performing it. For example, while food shopping for several items (on a list), can the patient demonstrate mental flexibility by holding one food item in his or her mind (e.g., milk) and still pick up cat food in the pet food aisle, or is the patient only able to seek one item at a time—even if he or she passes other food items on the shopping list?

Instruct the patient to perform two of the following tasks:

■ Simultaneously monitor the preparation of soup (in a pot on the stove) and toast in the toaster. Can the patient switch back and forth between two tasks involved in a meal preparation activity?

■ Iron several items of clothing while monitoring other clothing in the washer and dryer. Write checks to pay monthly bills while intermittently answering the phone or while monitoring the preparation of soup (in a pot on the stove). Can the patient switch back and forth between household chores?

- Simultaneously boil water for tea while completing a crossword puzzle, making a weekly shopping list, or reading a bus schedule. Can the patient switch back and forth between unrelated tasks?

☞ Cognitive impairment is indicated if the patient becomes confused while attempting to perform simultaneous tasks or forgets to complete or return to one task when engaged in a second task.

6. *Initiation of activity.* The patient should be able to initiate ADL independently or with minimal cues and set-up. Present the patient with the ingredients to prepare a peanut butter and jelly sandwich. State to the patient, "Show me what you would do with these items." Note whether the patient is able to initiate preparing the sandwich independently or with further cues. Left unattended, was the patient able to continue the task independently, or did the patient drift into an inactive state and fail to carry out the activity?

☞ Cognitive impairment is indicated if the patient does not initiate the task, or initiates the task but then drifts into an inactive state and fails to fully prepare the sandwich.

7. *Termination of activity.* Similarly, the patient should be able to terminate ADL independently or with minimal cues. Left unattended, the patient should be able to complete the above cold meal preparation independently or with minimal cues.

☞ Cognitive impairment is indicated if the patient is unable to terminate the task independently and instead uses the entire loaf of bread to make multiple sandwiches.

8. *LTM and STM.* Because LTM often is spared or returns as the patient recovers from neurologic insult, the patient with high-level cognitive functioning should demonstrate good LTM. Ask the patient to recall personal autobiographical information. For example, ask the patient at least four of the following questions:

- What year were you born?

- What are the name(s) of your siblings, spouse, and children?

- Where did you grow up?

- In what year did you graduate from school?

- In what year were you married?

☞ Cognitive impairment is indicated if the patient is unable to answer more than half of the asked questions accurately.

If impaired, STM often remains problematic for the patient with high-level neurological involvement. Ask the patient to recall events that occurred in the past 24 to 48 hours. For example, ask at least four of the following questions:

- What did you have for breakfast today?

- Did you have visitors last evening? Who were they?

- What therapy did you have today just before you came to occupational therapy?

- What is my name? Can you remember when you first met me?

- How many days or weeks have you been in the hospital?

☞ Cognitive impairment is indicated if the patient is unable to answer more than half of the asked questions accurately.

- Instruct the patient that you are going to ask him or her to remember three items (e.g., watch, necktie, coffee mug).

- State the three items and tell the patient that you are going to ask her to recall those items after a short time has passed.

- Go on to test the patient's fund of knowledge.

- After 3 to 5 minutes have passed, ask the patient to recall and name the three items. The ability to reproduce these items after a short interval indicates retentive memory span. Patients who demonstrate good memory can reproduce two or three items.

☞ Cognitive impairment may be indicated if the patient can reproduce only one (or no) item.

9. *New learning* is the ability to encode STM into LTM storage. New learning depends on good STM. Without the ability to transfer new information from STM to LTM storage, individuals have great difficulty learning new information. Patients often are unable to remember the names of individuals introduced after their brain damage has occurred. The names of therapists, hospital roommates, and newly born grandchildren often will not be encoded into LTM storage. Patients often will have difficulty remembering daily schedules, medication routines, and meal times without the aid of compensatory strategies (such as a memory or schedule book).

New learning is difficult to assess in a quick screening and often requires an in-depth evaluation for accurate assessment. To determine whether the patient can demonstrate new learning, ask him or her questions about new information learned after the injury occurred.

- Can the patient remember the names of familiar clinical staff members, for example, What is my name? What is the name of your doctor, nurse, and therapist? What is the name of your hospital or group home roommate(s)?

- Is the patient able to remember why he or she is in the hospital or clinical setting, for example, Do you remember the name of this facility? Do you remember why you are here?

☞ Cognitive impairment may be indicated if the patient is unable to answer more than half of these questions accurately.

To determine whether the patient can learn new tasks after the occurrence of brain damage:

- Teach the patient a new card game. Determine whether the patient is able to learn a new game and teach it to another patient.

- Teach the patient to use a new compensatory strategy (such as using a watch beeper to cue him or her to take medication at specific times). Tell the patient that whenever the watch beeper goes off, he or she must remember to take his or her medication. Set the watch beeper to go off several times during the screening procedures. Note whether the patient can remember that the watch beeper is a cue to take medication.

Note: Because new learning often requires a significant amount of repetition, it may be difficult to assess new learning in a quick screening. If the patient does not remember that the watch beeper is a signal to take medication, it does not necessarily mean that the patient is incapable of new learning. Rather, it may mean that more repetition is required before new learning can occur.

10. *Generalization of learning.* This concept is similar to new learning. Generalization of learning is the ability to transfer the skills needed for one task to a new task that is similar. Generalization of learning depends on an intact memory system. LTM must be intact for an individual to transfer learned skills already in LTM storage to a new but similar situation. STM also must be intact for the individual to learn the skills needed for the new, similar task.

The individual must recognize the similarity between the old task and the new task. This requires the cognitive skills of *comparison, analysis, discrimination,* and the ability to *determine relationships between situations.* Often, when STM is impaired, individuals with brain damage have great difficulty generalizing already learned information to a novel task. If the patient demonstrates poor new learning, generalization of learning also will likely be poor.

Can the patient transfer the skills already learned for laundry using the laundry machine in his or her former apartment building to the skills needed to operate a similar but different laundry machine in his or her new community assisted living residence or the occupational therapy kitchen? Use the following steps to determine transfer of skills:

- Demonstrate how to use the washing machine in the occupational therapy kitchen (or community group home if that is the site in which the screening is being carried out).

- Ask the patient to wash a load of laundry using the novel washing machine.

Similarly, can the patient transfer the skills already learned for hot meal preparation using the microwave oven in his or her former apartment building to use of a different microwave oven in his or her new community assisted living residence or the occupational therapy kitchen?

- Demonstrate how to use the microwave oven in the occupational therapy kitchen (or community group home).

- Ask the patient to make tea using the novel microwave.

☞ Cognitive impairment may be indicated if the patient is unable to generalize existing knowledge to newly learned situations after significant repetition.

Note: Generalization of learning is another skill that is difficult to assess with a quick screening procedure. Because generalization of learning often requires a significant amount of repetition and practice, an in-depth evaluation often is needed to assess generalization of learning accurately. If a patient is unable to use a novel washing machine or microwave oven, it does not mean that generalization of learning did not occur. Rather, it indicates that further opportunities for learning are needed.

11. *Fund of knowledge.* The patient should be able to demonstrate a good to normal level of fund of knowledge largely because fund of knowledge depends on an intact LTM system. Ask the patient at least four of the following questions:

- What is the name of the current president? The first president? The president before the current president?

- What is the name of a large city in the northeastern United States?

- How many weeks are in a year? How many days are in a year?

- Name as many oceans and bodies of water as possible.

- Name as many states, countries, and continents as possible.

☞ Cognitive impairment is indicated if the patient is unable to answer more than half of the questions accurately.

12. *Sequencing.* Ask the patient to perform at least three of the following tasks:

- Count forward and backward (e.g., from 27 to 37).

- Count forward and backward by 7s (e.g., 7, 14, 21, 28). This sequencing skill also requires *calculation* (a function of the left frontal lobe).

- State the months of the year in sequence—forward and backward.

- Describe the steps of laundry (or fixing a flat tire, preparing a peanut butter and jelly sandwich).

- Demonstrate the steps of ordering pizza from a pizza shop using the telephone, phone book, paper, and pencil provided.

☛ Cognitive impairment is indicated if the patient inaccurately sequences two of the three tasks.

13. *Categorization.* Ask the patient to name as many items as possible in a specific category. Ask at least three of the following:

- Name as many farm animals as possible.

- Name as many types of automobile makers as possible.

- Name as many modes of transportation as possible.

- Name as many words as possible that begin with the letter *P.*

☛ Most adults can name 20 to 25 items in a single category within 1 minute. Cognitive impairment is indicated by an inability to generate more than 3 to 5 items in each of the three indicated categories.

The ability to cognitively generate items belonging to a specific category is a test of concept formation and thought generation. Ask the patient to sort items into separate piles. Sorting items into same-category piles depends on *visual attention* and *visual discrimination,* the ability to distinguish similar but different items. Ask the patient to perform at least two of the following sorting tasks:

- Sort utensils into groups of spoons, forks, and knives.

- Sort cookies into their different types.

- Sort nuts and bolts into piles of different sizes.

- Sort different spools of thread by their color (or size).

☛ Cognitive impairment may be indicated if the patient makes more than two or three errors in each of the two sorting tasks. It is important to distinguish between an impairment in categorization and an impairment in visual discrimination.

14. *Command following.* The patient should be able to follow multiple-step commands with minimal difficulty. Ask the patient to perform at least one of the following:

- Prepare a cold meal using written and verbal multiple-step directions.

- Fill out a form with multiple written steps (e.g., a W4 form for work).

- Follow verbal and/or written directions as appropriate with regard to the patient's deficits.

☛ Cognitive impairment is indicated if the patient is unable to follow directions without making two or three errors in one task.

15. *Insight* is the ability to demonstrate an understanding of one's own strengths and weaknesses, motivations, and behaviors. Whereas the patient with low-level cognitive functioning often lacks insight into his or her strengths and weaknesses, motivations, and behaviors, the patient with high-level cognitive functioning often is able to acknowledge deficit areas and use appropriate strategies to compensate. Ask the patient to describe both personal strengths and weaknesses. Note whether the patient is aware of deficit areas. Ask the patient to describe how his or her deficits have affected daily life skills.

16. *Safety/judgment* largely depends on an individual's insight or awareness of deficit areas. Lack of insight is likely if a patient demonstrates poor safety/judgment. Patients who are not aware of their own deficit areas will likely be a safety risk to themselves and to others. Is the patient a safety risk to him- or herself and/or others? For example, is the patient unaware that his or her hemiplegic arm is hanging over the wheelchair armrest, in the wheelchair spokes? Is the patient impulsive? Impulsivity often results from orbitofrontal lobe damage. For example, does the patient attempt to get out of the wheelchair without locking the brakes and moving the footplates away? Does the patient reach for hot items on the stove without using pot holders? Does the patient attempt to walk out in traffic without observing the traffic signals? Does the patient understand simple emergency procedures to use if an accident occurs? Describe at least four of the following emergency situations and ask the patient to determine what he or she would do:

- You are asleep in your room at home. The smoke alarm goes off. What should you do?

- You lose your keys and are locked out of your house. What should you do?

- You are home alone. You fall and are unable to move your leg. You are in great pain. What's the best thing to do?

- You are preparing dinner. High flames begin to come from the pan on the stove. What should you do?

- You are in a large shopping mall and become separated from the person you came with. You are lost. What should you do?

☞ Safety may be compromised if the patient demonstrates any of the above behaviors or is unable to offer appropriate solutions for any of the indicated emergency situations.

17. *Planning* is the ability to mentally conceptualize an event or activity that will occur in the future. Planning involves the ability to assess task demands, analyze and synchronize the components of a task, consider a range of options, and make a decision regarding the best course of action.

Planning requires the process of *abstraction,* which is an ability thought to be mediated by the right frontal lobe. Often when the right frontal lobe is damaged, patients display *concrete thought,* which is a tendency to interpret events and language by their literal meaning.

A patient with high-level cognitive functioning should be capable of planning to a moderate extent. Ask the patient to perform at least one of the following:

- Plan a shopping list for an upcoming barbecue for 20 people.

- Plan a weekly food shopping list for yourself.

- Plan a monthly budget for yourself. Note whether the patient is able to consider monthly bills, paycheck or social security income, recreational expenses, transportation expenses, and food costs.

- Plan a weekend vacation to the shore.

- Plan a list of items you will need to store in your home in case an accident or emergency occurs. What types of information or phone numbers do you need?

☞ Cognitive impairment may be indicated if the patient demonstrates difficulty planning a basic self-care task of two or three steps.

18. *Problem solving* is a high-level skill involving planning and abstraction. If planning and abstraction are impaired, the patient will likely display difficulty with problem solving. Planning involves the following:

- Problem recognition—Can the patient recognize that a problem exists?

- Problem analysis—Is the patient able to understand the causes of a problem?

- Problem resolution—Is the patient able to consider a range of possible resolutions? Is the patient able to select the best option for resolution? Is the patient able to implement the selected resolution?

- Resolution assessment—Is the patient able to assess how well the selected resolution worked?

- Resolution revision—Is the patient able to revise the resolution plan if the first attempted resolution did not work optimally?

Ask the patient to problem solve at least one of the following scenarios:

- You get on the bus to go to work but find that you have taken the wrong bus and have ended up at a destination 10 miles from your work site. You are now 45 minutes late for work. What should you do? What are your options?

- Your roommate smokes in bed at night before going to sleep. You realize that this causes a safety risk to both of you. What could you do? What are your options?

■ One night there is a thunderstorm that knocks out the electrical power to your home for several hours. Your phone lines also are lost. You see that your neighbor's tree has been struck by lightning and is on fire. What should you do?

■ You realize that you miscalculated your monthly budget and have found yourself at mid-month with enough money to buy only groceries. You still need to pay monthly bills and pay for your transportation to work. What can you do? What are your options?

☛ Cognitive impairment is indicated if the patient is unable to offer an appropriate resolution to any of these situations.

19. *Self-control* is the ability to monitor and regulate one's impulses, urges, and compulsions and is considered to be one of the executive functions—the cognitive skills one needs to override socially unacceptable urges in order to behave in socially appropriate ways. It requires the patient to possess insight regarding deficit areas consider the consequences of his behaviors

Disinhibition—or loss of the ability to self-regulate socially inappropriate behaviors—often occurs as a result of orbitofrontal lobe damage. The patient may display impulsivity, aggression, irritability, agitation, and/or sexual disinhibition. During the screening procedures the therapist should note whether the patient presents any of the following behaviors of disinhibition:

■ Does the patient verbalize inappropriate sexually related comments?

■ Does the patient act impulsively? Does the patient act on any impulse that enters his or her consciousness?

■ Is it apparent that the patient acts on impulse without considering the consequences of his or her behavior? Is the patient able to consider how his or her behaviors will affect others?

■ Is the patient verbally or physically aggressive?

■ Does the patient present irritability or agitation in response to the slightest provocation?

■ Is the patient labile? Does he or she have difficulty controlling his or her emotions? Is the patient frequently tearful? Do the patient's moods change suddenly, without correspondence to environmental events?

☛ Impaired executive functions mediating self-control are indicated by demonstration of any of the above behaviors.

RED FLAGS
SIGNS AND SYMPTOMS OF COGNITIVE IMPAIRMENT

A red flag is an indicator or symptom of dysfunction. If a patient is observed displaying any of the following red flags, it is likely that the patient possesses cognitive deficits.

- **CONFUSION/DISORIENTATION** is demonstrated by the patient's lack of understanding of present events and disorientation regarding where the patient is, why the patient is in the hospital, and what month or season it is.

- **CONFABULATION** is the generation of false information to account for memories that the patient is unable to recall. It is not uncommon for patients with brain damage to generate intricate and complex false stories in an attempt to fill in missing parts of their memory.

- **CONCRETE THINKING** occurs when patients are unable to interpret events and language with any meaning other than a literal one. Often jokes are lost on patients with concrete thinking because they interpret the language of the joke literally rather than figuratively. Innuendoes and symbolic meanings are often lost on individuals who demonstrate concrete thinking.

- **DELAYED PROCESSING TIME** is demonstrated by the patient's inability to answer questions or formulate ideas in a timely manner. Instead, a delay occurs in the patient's mentation. Thus, it is important to give the cognitively involved patient sufficient time to answer questions and formulate ideas.

- **DISINHIBITION** occurs when the patient is unable to monitor and regulate socially inappropriate impulses and behaviors. Often the disinhibited patient verbalizes sexualized language and may dress and behave in socially inappropriate ways. Disinhibition is sometimes demonstrated by the act of removing one's own clothing in public and making sexual propositions to both familiar others and strangers.

- **DISTRACTIBILITY** often is displayed in high levels by patients with brain damage. These patients have difficulty remaining on one task for any length of time. External verbal cues often are needed to help the patient attend to one task at a time.

- **TANGENTIAL SPEECH, OR FLIGHT OF IDEAS,** occurs when patients are unable to concentrate on one idea at a time for any length of time. Instead, patients jump from thought to thought, often without any obvious connection between thoughts. Their verbalizations appear to be a stream of unrelated ideas.

- **PERSEVERATION** is the inability to stop an activity once started. Patients are unable to interpret cues that indicate they need to stop the task or change strategies. Instead they continue to implement the behavior over and over. For example, a patient who is attempting to load the dishwasher but finds that he or she cannot fit all of the dishes may continuously rearrange the dirty dishes in a futile attempt to make all of the dishes fit in the dishwasher.

- **MEMORY DEFICITS,** particularly STM deficits, often indicate neurologic impairment. STM deficits that worsen over time may indicate dementia, brain tumor, or other neurological pathology.

- **POOR INSIGHT** is the lack of an accurate awareness of one's strengths and deficit areas relating to one's functional status. As a result of poor insight, patients commonly attempt activities that are at too high a level, often causing themselves a series of failure experiences. Patients also are unable to draw relationships between their own behaviors and others' responses to the patient's behaviors.

- **POOR SAFETY JUDGMENT** is the inability to discern the inherent danger in a specific situation. Consequently, patients commonly become involved in situations in which they place themselves at risk for injury or assault and unwittingly allow themselves to be taken advantage of.

AVAILABLE IN-DEPTH ASSESSMENTS

If cognitive screening reveals a neurological impairment, the following in-depth assessments are available:

Allen Cognitive Level Test (ACL)
C. K. Allen
Any cognitively or psychiatrically impaired population
S&S Worldwide
PO Box 513
Colchester, CT 06415-0513
(800) 243-9232

Allen Diagnostic Module (ADM)
C. A. Earhart, C. K. Allen, T. Blue
Adolescents and adults at Allen Cognitive Level 3
S&S Worldwide
PO Box 513
Colchester, CT 06415-0513
(800) 243-9232

Arizona Battery for Communication Disorders of Dementia (ABCD)
K. Bayles, C. K. Tomoeda
Adults with Alzheimer's disease
Canyonlands Publishing, Inc.
141 South Park Ave.
Tucson, AZ 85719

Arnadóttir OT-ADL Neurobehavioral Evaluation (A-ONE)
G. Arnadóttir
Adults with cortical disorders
Arnadóttir, G. (1990). *The brain and behavior: Assessing cortical dysfunction through activities of daily living.* Philadelphia: Mosby.

Assessment of Motor and Process Skills (AMPS)
A. G. Fisher
Children and adults with developmental, neurologic, and/or musculoskeletal disorders
Anne G. Fischer, ScD, OTR/L, FAOTA
AMPS Project
Occupational Therapy Building
Colorado State University
Fort Collins, CO 80523

Autobiographical Memory Interview
M. Kopelman, B. Wilson, A. Baddeley
Organic amnesic syndromes, dementia, psychiatric disorders
National Rehabilitation Services
117 North Elm St.
PO Box 1247
Gaylord, MI 49735
(517) 732-3866

Bay Area Functional Performance Evaluation (2nd ed.) (BaFPE)
S. L. Williams, J. Bloomer
Adults with psychiatric and neurologic impairment
Maddack, Inc.
6 Industrial Rd.
Pequannock, NJ 07440
(800) 443-4926

Blessed Dementia Rating Scale
G. Blessed, B. E. Tomlinson, M. Roth
Elderly persons with dementia
Blessed, G., Tomlinson, B. E., & Roth, M. (1968). The association between quantitative measures of dementia and of senile change in the cerebral

gray matter of elderly subjects. *British Journal of Psychiatry, 114,* 797–811.

Burns Brief Inventory of Communication and Cognition (BBRS)

M. S. Burns
Adults with neurologic impairment
Psychological Corporation
555 Academic Ct.
San Antonio, TX 78204-2498.
(800) 228-0752, (800) 211-8378

Canadian Occupational Performance Measure (COPM)

M. Law, S. Baptiste, A. Carswell, M. A. McColl, H. Polatajko, N. Pollock
Children and adults with cognitive, psychosocial, neurologic, and/or musculoskeletal impairment
Canadian Association of Occupational Therapists
110 Eglington Ave. West, 3rd Floor
Toronto, ON M4R 1A3 Canada

Clinical Dementia Rating (CDR)

J. C. Morris
Adults with dementia of the Alzheimer's type
John C. Morris, MD
Memory and Aging Project
Washington University School of Medicine
660 South Euclid Ave.
PO Box 8111
St. Louis, MO 63110

Clock Test

H. Tuokko, T. Hadjistravropoulos, J. A. Miller, A. Horton, B. L. Beattie
Elderly persons with dementia
Multi-Health Systems, Inc.
908 Niagara Falls Blvd.
North Tonawanda, NY 14120-2060
(416) 424-1700

Cognitive Adaptive Skills Evaluation (CASE)

Adults with psychiatric disorders and adolescents and adults with cognitive impairment
G. N. Masagatani, C. S. Nielson, E. R. Ranslow, Gladys Masagatani, Ed, OTR, FAOTA
Eastern Kentucky University
Department of Occupational Therapy

Dizney 103
Richmond, KY 40475-3135

Cognitive Behavior Rating Scales (CBRS)

J. M. Williams
Adults with neurologic impairment
Psychological Assessment Resources, Inc.
PO Box 998
Odessa, FL 33556-9908
(800) 727-9329

Cognitive Performance Test (CPT)

T. Burns
Adults with Alzheimer's disease
Burns, T., Mortimer, J. A., & Merchak, P. (1994). Cognitive Performance Test: A new approach to functional assessment in Alzheimer's disease. *Journal of Geriatric Psychiatry and Neurology, 7,* 46–54.

Contextual Memory Test (CMT)

J. P. Toglia
Adults with memory deficits secondary to acquired or organic neurologic impairment
Therapy Skill Builders
Psychological Corporation
555 Academic Ct.
San Antonio, TX 78204-2498
(800) 228-0752, (800) 211-8378

Dementia Rating Scale (DRS)

S. Mattis
Adults with dementia
Psychological Assessment Resources, Inc.
PO Box 998
Odessa, FL 33556-9908
(800) 331-8378

Elemental Driving Simulator (EDS) and Driving Assessment System (DAS)

R. Giatnutsos, A. Campbell, A. Beattie, F. Mandriota
Adults with neurologic disorders
Life Sciences Associates
One Fenimore Rd.
Bayport, NY 11795-2115
(516) 472-2111

Global Deterioration Scale (GDS), including Brief Cognitive Rating Scale (BCRS) and Functional Assessment Staging (FAST)
GDS: B. Reisberg, S. H. Ferris, M. J. De Leon, T. Crook
BCRS & FAST: B. Reisberg, S. G. Sclan, E. Franssen, A. Kluger, S. Ferris
Adults with primary degenerative dementia
Reisberg, B., Sclan, S. G., Franssen, E., Klugar, A., & Ferris, S. (1994). Dementia staging in chronic care populations. *Alzheimer Disease and Associated Disorders, 8*(Suppl. 1), S188–S205.
Reisberg, B., & Ferris, S. H. (1988). The Brief Cognitive Rating Scale (BCRS). *Psychopharamacology Bulletin, 24,* 629–636.
Barry Reisberg, MD
Department of Psychiatry
New York University Medical Center
550 First Ave.
New York, NY 10016

Learning and Memory Battery (LAMB)
J. P. Schmidt, T. N. Tombaugh
Adults with memory impairment
Multi-Health Systems, Inc.
908 Niagara Falls Blvd.
North Tonawanda, NY 14120-2060
(416) 424-1700

Katz Index of ADL
S. Katz, A. B. Ford, R. W. Moskowitz, B. A. Jackson, M. W. Jaffe
Elderly persons with neurologic and/or musculoskeletal impairment
Katz, S., Ford, A. B., Moskowitz, R. W., Jackson, B. A., & Jaffe, M. W. (1963). The Index of ADL: A standardized measure of biological and psychological function. *Journal of the American Medical Association, 185,* 914–919.

Kitchen Task Assessment (KTA)
C. Baum, D. F. Edwards
Adults with dementia
Baum, C., & Edwards, D. F. (1993). Cognitive performance in senile dementia of the Alzheimer's type: The Kitchen Task Assessment. *American Journal of Occupational Therapy, 47,* 431–436.

Klein–Bell Activities of Daily Living Scale
R. M. Klein, B. J. Bell
Infants to adulthood neurologic and/or musculoskeletal impairment
Health Sciences Center for Educational Resources
University of Washington
T-281 Health Science Bldg.
Box 357161
Seattle, WA 98195-7161

Kohlman Evaluation of Living Skills (KELS)
L. Kohlman-Thomas
Adolescents and adults with psychiatric, developmental, and/or neurologic impairment
American Occupational Therapy Association
4720 Montgomery Ln.
PO Box 31220
Bethesda, MD 20824-1220
(301) 652-2682

Leiter International Performance Scale–Revised
G. H. Roid, L. J. Miller
Children, adolescents, and adults with cognitive impairment
Stoelting Company
620 Wheat Ln.
Wood Dale, IL 60191
(800) 860-9775

Lowenstein Occupational Therapy Cognitive Assessment (LOTCA)
M. Itzkovich, B. Elazar, S. Averbuch, N. Katz, Lowenstein Rehabilitation Hospital, Israel
Adults with neurologic impairment
Maddak, Inc.
6 Industrial Rd.
Pequannock, NJ 07440
(800) 443-4926

Rivermead Behavioural Memory Test (RBMT)
B. Wilson, J. Cockburn, A. Baddeley
Children, adolescents, and adults with memory impairment
National Rehabilitation Services
117 North Elm St.
PO Box 1247

Gaylord, MI 49735
(517) 732-3866

Routine Task Inventory (RTI–2)
C. K. Allen
Adults with cognitive impairment
Allen, C. K., Earhart, C. A., & Blue, T. (1992).
*Occupational therapy treatment goals for the
physically and cognitively disabled.* Bethesda, MD:
American Occupational Therapy Association.

**Scales of Cognitive Ability for Traumatic
Brain Injury (SCATBI)**
B. Adamovich, J. Henderson
Adults with traumatic brain injury
Applied Symbolix
800 North Wells St., Ste. 200
Chicago, IL 60610

Severe Cognitive Impairment Profile (SCIP)
G. M. Peavy
Adults with severe dementia
Psychological Assessment Resources, Inc.
PO Box 998
Odessa, FL 33556-9908
(800) 331-8378

Severe Impairment Battery (SIB)
Elderly with severe, end-stage dementia
J. Saxton, K. L. McGonigle, A. A. Swihart,
F. Boller
National Rehabilitation Services
117 North Elm St.
PO Box 1247
Gaylord, MI 49735
(517) 732-3866

**Structured Observational Test of Function
(SOTOF)**
A. Laver, G. E. Powell
*Elderly persons with neurologic impairment or
dementia*
NFER-NELSON Publishing Company
Darville House
2 Oxford Rd. East
Windsor, Berkshire SL4 1DF England
0753 858961

Test of Everyday Attention (TEA)
I. H. Robertson, T. Ward, V. Ridgeway,
I. Nimmo-Smith
*Adults with cognitive attentional deficits secondary
to neurologic impairment*
National Rehabilitation Services
117 North Elm St.
PO Box 1247
Gaylord, MI 49735
(517) 732-3866

**Test of Orientation for Rehabilitation Patients
(TORP)**
J. Dietz, C. Beeman, D. Thorn
*Adolescents and adults with confusion and
disorientation secondary to neurologic impairment;
used only in inpatient settings with individuals who
have been in treatment for a minimum of 1 week*
Therapy Skill Builders
Psychological Corporation
555 Academic Ct.
San Antonio, TX 78204-2498
(800) 228-0752, (800) 211-8378

For further information about the these
assessments, refer to the following resources:

Asher, I. E. (1996). *Occupational therapy
assessment tools: An annotated index* (2nd ed.).
Bethesda, MD: American Occupational
Therapy Association.
Impara, J. C., Plake, B. S., & Murphy L. L. (Eds.).
(1998). *The thirteenth mental measurements
yearbook.* Lincoln: University of Nebraska
Press.
Murphy, L. L., Impara, J. C., & Plake, B. S. (Eds.).
(1999). *Tests in print: An index to tests, test
reviews, and the literature on specific tests* (Vols.
1 & 2). Lincoln: University of Nebraska Press.
Neistadt, M. E. (2000). *Occupational therapy
evaluation for adults: A pocket guide.*
Philadelphia: Lippincott, Williams & Wilkins.
Plake, B. S., Impara, J. C., & Murphy, L. L. (Eds.).
(1999). *The supplement to* The Thirteenth
Mental Measurements Yearbook. Lincoln:
University of Nebraska Press.

COGNITIVE SCREENING: LOW LEVEL

Patient Name _____   Date _____

Cognitive Skill	No Impairment Detected	Possible Impairment Detected	Comments
Alertness/arousal			
Orientation			
Recognition			
Attention/concentration			
Initiation of activity			
Termination of activity			
Memory LTM			
STM			
Fund of knowledge			
Sequencing			
Categorization			
Command following			
Safety/judgment			
Self-control			

Note. From *Levels of Cognitive Functioning,* by C. Hagen, 1980, Downey, CA: Rancho Los Amigos Medical Center, Adult Brain Injury Service. Copyright 1980 by C. Hagen. Reprinted with permission.

COGNITIVE SCREENING: HIGH LEVEL

Patient Name _____ Date _____

Cognitive Skill	No Impairment Detected	Possible Impairment Detected	Comments
Alertness/arousal			
Orientation			
Recognition			
Attention/concentration Mental flexibility			
Initiation of activity			
Termination of activity			

Memory
    LTM
    STM			

New learning			
Generalization of learning			
Fund of knowledge			
Sequencing			
Categorization			
Direction following			
Insight			
Safety/judgment			
Planning			
Problem solving			
Self-control			

Note. From *Levels of Cognitive Functioning,* by C. Hagen, 1980, Downey, CA: Rancho Los Amigos Medical Center, Adult Brain Injury Service. Copyright 1980 by C. Hagen. Reprinted with permission.

Visual
SCREENING

FUNCTIONAL IMPLICATIONS OF VISUAL IMPAIRMENT

The visual system is our primary sensory system. We use vision to identify, perceive, and interpret all objects and situations we encounter in our environment. The visual system is closely linked with our motor/postural and vestibular systems, which enable us to motorically plan movements, move within our environment, and maintain an upright position in space. According to Warren (1997), the visual system allows us to instantly, accurately, and reliably attend to environmental information, integrate it, and use it to make daily decisions. The visual system allows us to respond to a situation by interpreting nonverbal body language and environmental dangers. The visual system is essential for learning.

The visual system consists of two systems: the *anterior visual system* and the *posterior visual system*. The anterior visual system includes all structures anterior to the optic chiasm; this includes the cornea, iris, pupil, lens, aqueous and vitreous humor, retina, and optic nerve. The main goal of the anterior system is to provide high-quality, accurate input. These structures allow light to be refracted via different structural and fluidlike mediums and projected onto the back of the eye, or the retina. According to Warren (1997), the mechanics of these structures can be compared to a camera in that a picture is taken (lens) and projected onto the retina (film).

The posterior visual system includes the optic chiasm, optic tracts, lateral geniculate nucleus, superior/inferior colliculi, geniculocalcarine tracts, and the occipital cortex. This system is responsible for the transferring of information from the anterior system to the posterior system, where it is then processed, interpreted, and used to make daily choices and decisions.

The *cortical processing centers*—which include the *temporal* and *parietal circuits* and *prefrontal* and *medial temporal lobes*—assist in processing visual information, allowing for adaptation to the environment. The *brain stem* is responsible for pupillary responses and reflexes working in conjunction with the cerebellum for control of eye movements and coordination of the entire visual system. Together this triad of the anterior, posterior, and cortical systems allow for functional vision. All components of the anterior, posterior, and cortical processing areas compliment one another and must be in continuous communication.

Visual impairment can occur secondary to illness, trauma, and age by altering the quality and quantity of visual input into the central nervous system (CNS) and how the CNS uses incoming visual information. Impairments in visual acuity might be caused by disease process, injury, trauma, or retinal artery stroke. Age-related diseases such as *macular degeneration, diabetic retinopathy, glaucoma,* and *cortical blindness* all affect a person's central and/or peripheral vision, acuity, contrast sensitivity, and ability to detect motion. A patient with reduced visual acuity due to macular degeneration, diabetic retinopathy, or cortical blindness may exhibit increased difficulty with near tasks, such as reading newsprint, bills, and nutrition and medication labels. This condition is due to areas of blindness called *scotomas*. Patients with *peripheral field loss* (tunnel vision) due to glaucoma may exhibit difficulties detecting movement or objects within their periphery. Tasks such as navigating outside within dynamic environments, crossing streets, driving, or shopping in a grocery store present great challenges.

Clothing selection presents a significant challenge, as color discrimination among blues, blacks, and browns is severely impaired due to involvement of the cones in the macular region. Patients experience difficulties with low-contrast environments and tasks. For example, a person with macular degeneration may present with feeding difficulties when pouring black coffee into a dark mug or eating pasta with a white sauce on a white plate. Navigational difficulties and increased number of falls are exhibited with unmarked steps of the same color, negotiating a curb or a slope in the sidewalk, and sorting white laundry. The inability to recognize a person may be confused with cognitive deficits because the face is a low-contrast feature. A simple task such as walking outside on a bright, sunny day to a mailbox, then returning to a dimly lit home may increase a person's risk of falls due to increased *glare sensitivity* and the inability to accommodate to changing lighting.

Injury due to blunt trauma, facial fracture, and nerve stretching to the anterior portion of the visual system will yield *refractive errors*. Patients with refractive errors may complain of blurry print or increased difficulty when reading (*hyperopia*) or increased difficulty reading street signs at a distance (*myopia*). Patients with visual field deficits will experience impairment in visual scanning/exploration, saccades, pursuits, reading, writing, mobility, and all aspects of activities of daily living (ADL) and instrumental activities of daily living tasks.

Functional mobility is severely affected by visual field deficits. Patients present with slower gait, shorter strides, anticipation of movements, shoe gazing, and tactile strategies such as trailing a wall with their finger during ambulation. Patients with *parietal tract* involvement will experience increased difficulty with steps or changing support surfaces as their inferior visual field is affected. Patients with involvement of the *temporal tract* will demonstrate difficulties with their superior visual field. For example, identifying traffic lights when driving, navigating from wheelchair level, or scanning the top shelves in a grocery store will be most challenging. Patients with right or left *visual field cuts* may experience increased collisions secondary to an inability to see the visual information in that part of their visual field.

Reading tasks are affected due to slower speeds, transpositions, or omissions of ototype (print size), abbreviated scanning, and line skipping and jumping. A patient with a visual field deficit may have difficulty addressing envelopes, writing checks, and filling out forms. These patients will demonstrate problems with word positioning and spacing as well as difficulty writing on a straight line.

Patients will experience increased difficulty in dynamic environments more than static environments. These patients typically stay home, become socially isolated, and dislike having items moved within their home.

The *cortical areas of the visual system* house the nuclei for *cranial nerves 3, 4, and 6,* which control the extraocular muscles of the eye, the neural control systems (which enable *foveation*— the ability to keep one's eye on the target), gaze stabilization, and postural and motor control.

The extraocular muscles of cranial nerves 3, 4, and 6 mediate eyeball movements, eyelid opening, pupillary size, and binocular function. Pathology in any of these cranial nerves will usually cause *vertical* or *horizontal diplopia* due to a *strabismus* and impaired *visual scanning*.

Persons with *cranial nerve 3* involvement may demonstrate difficulty with near tasks such as reading, buttoning, or medication identification. Patients may also present with *ptosis* (drooping) of the involved eyelid, and patients will complain of horizontal diplopia. Persons with *cranial nerve 4* involvement will complain of vertical diplopia in all gazes. They will have difficulty on near and far tasks. Persons with *cranial nerve 6* involvement will complain of horizontal diplopia at far distances secondary to the involvement of the lateral rectus. Patients may have difficulty watching television, telling time from across the room, and in performing tasks requiring functional mobility.

Patients will have difficulty with accommodation, fixation, smooth pursuit, binocular fusion, and saccades, affecting all ADL tasks by requiring increased head turning to compensate for limited eye movements. Patients with *eye movement disorders* such as fixation, pursuit, and saccades have difficulty reading or following cars in traffic. Patients with accommodative disorders have difficulty adjusting vision between near and far distances. For example, a student may have difficulty reading the blackboard and taking notes. The *vestibular* as well as the *ocular* and *optokinetic reflexes,* which help to hold images on the retina during stable and motionary head movements, may be impaired. These reflexes enable one to ride in a car and read highway signs without blurring. Patients with *cerebellar findings* will experience jerky eye movements and nystagmus, making fixation difficult. Functionally, patients may be unable to stabilize their gaze during both near and far tasks.

FUNCTIONAL VISUAL SCREENING PROCEDURES

Visual screenings should be performed on all individuals with traumatic brain injury, acquired brain injury, neuromuscular deficits, diabetes, glaucoma, macular degeneration, a history of falls, and complaints of visual changes (e.g., blurry, double, or hazy vision) as well as for patients over 65 years of age. It is important to conduct a thorough chart review because vital information can be ascertained, for example,

- *Diagnosis.* If a patient has multiple sclerosis, one might anticipate optic neuritis, which would affect functional vision.

- *Lesion/injury.* If a patient sustained an occipital cerebrovascular accident, one might anticipate cortical blindness.

- *Surgical history.* If a patient has had a temporal lobe brain tumor resected, one might suspect a visual inattention or visual field cut.

- *Medical history.* If a patient has a history of diabetes, one might anticipate diabetic retinopathy.

- *Medication.* A patient who is on seizure prophylaxis may complain of blurry/double vision.

- *History of falls.* The therapist should question whether the falls relate to balance or visual deficits.

- *Recent consults.* These may provide information from an ophthalmology consultation.

- *Other discipline notes.* This information may reveal physical therapy's report of patient colliding on left side during ambulation.

A basic cognitive screening should be considered before the functional visual evaluation. A therapist should also determine whether the patient is capable of accurate yes/no responses. Because of the increased likelihood of visual field deficits among patients with brain injury, evaluations can be broken up into shorter segments to maximize sustained attention and reliable responses. Therapists should practice alternative communication techniques for patients with expressive aphasia, such as communicating with pointing, head nodding, or eye blinking. It is important to interview both the patient and the family during the patient history taking. Although patients with high-level cognitive functioning may be able to participate and answer questions, they may not always be forthcoming with information due to denial, embarrassment, or cognitive deficits. Results and interpretation of the functional visual screening should emphasize not how much the results deviate from the norm but how the results affect ADL performance.

Visual screening should be conducted in a low-stimuli room without distractions. Rooms should have the capacity to adjust lighting (either very dark or brightly illuminated). When performing the functional mobility screen on a patient, always check the patient's ambulation status, guarding, and appropriate assistive device used, if any. The therapist should carefully guard the patient to eliminate risk of falls or injury. Observation of patient performance and behavior should begin immediately.

Functional visual screening allows the therapist to determine whether a deficit area exists and whether further in-depth visual evaluation is required. Patients should always wear corrective lenses during exam. Visual screening should be separated for patients with low-level and high-level cognitive functioning. A visual screening for a low-level/comatose patient should include the following:

- Visual history

- Observation

- Acuity

- Pupillary size

- Ocular alignment

- Object fixation

- Visual pursuit.

In addition to the basic visual skill evaluation for the low-level/comatose patient, a visual screening for a high-level patient should include

- Acuity

- Extraocular motility

- Diplopia testing

- Saccade

- Accommodation

- Convergence

- Visual fields

- Visual acuity

- Contrast sensitivity

- Visual skills during functional mobility.

LOW-LEVEL/COMATOSE PATIENT

When assessing a patient who is comatose or minimally conscious, most information is derived from the medical charts and patient observation. Because the patient is unable or minimally able to participate in the visual evaluation, it is important to establish the presence of sight, although at this time it is impossible to quantify the level of acuity. It is important to reassess patients every few days. As their arousal level improves, it will allow the therapist to begin assessing more basic visual skills.

It is best to position the patient upright if possible, either in bed (with the head of the bed elevated) or seated in a supportive wheelchair with head control. This will maximize the patient's level of arousal and allow for easier positioning during screening administration.

When screening the low-level/comatose patient for visual function, the following procedures are used:

1. *Visual history.*

- What is the location of the lesion/injury or infarct?

- Does the patient have a history of eye trauma, surgery, or an eye-related ocular disease such as diabetes, macular degeneration, glaucoma, or cataracts?

- Does the patient wear glasses or corrective lenses? Does the patient have a history of falls?

Note: Information should be taken from the medical chart as well as from family members. Asking these questions allows the therapist to anticipate potential visual dysfunction and findings. Additionally, as the patient begins to emerge from a low-level state, this information will assist in further evaluation.

2. *Observation.*

- Is the patient opening his or her eyes? If yes, are both eyes equal?

- Is there a difference in eye opening with the room lights on or off?

- Can the patient fixate on something or someone?

- Can the patient track a bright object? If yes, are both eyes moving together?

- Does the patient demonstrate a gaze preference to one side?

- Does one eye deviate in or out?

- Is the patient blinking excessively?

Note: If yes is indicated in any of these questions, further visual screenings should be performed as the patient begins to demonstrate increased arousal.

3. *Acuity* is the ability to see visual detail. Use the following tests for optokinetic response:

- In a well-illuminated room with minimal distractions, present the patient with a high-contrast object (a white index card with a 2-inch black square drawn on it), holding it approximately 16 inches from the patient's eyes.

- Test one eye at a time by covering the nontested eye with a patch or gauze pad.

- Quickly move target right to left.

☞ Nystagmus should be present, indicating preliminary sight. Impairment may be indicative of optic nerve atrophy (cranial nerve 2), occipital lobe damage, vitreous hemorrhage, or macular scotoma.

Note: When a patient is comatose or has low-level cognitive functioning, determination of visual acuity is made very grossly. Exact acuity level is impossible to ascertain; however, nystagmus is indicative that the patient is able to perceive the stimulus and has basic sight.

4. *Pupillary size* is controlled by the brain stem and cranial nerve 3. Pupillary constriction and dilation controls the amount of light entering the eye. Functionally, pupillary constriction allows us to adapt to illumination changes within the environment.

- Observe the pupil size of each eye in a well-lit room.

- Compare the size of both pupils.

- Note size in millimeters. Normal pupil size is 3 mm in both eyes.

- In a dimly lit room, hold a penlight approximately 4 inches from the patient's eye.

- Shine light into one eye for 2 seconds (see Figure 2.1).

- Observe for a rapid constriction and maintenance of stimulated pupil.

- Observe for equal constriction of nonstimulated pupil. Normal pupillary constriction from a light source is 1.5 mm in both eyes.

- Repeat on nonstimulated eye.

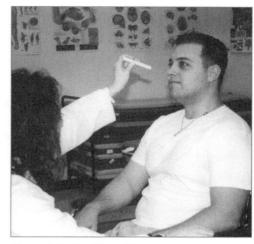

Figure 2.1 Pupillary Size

☞ A visual impairment is indicated if the pupil is greater or smaller than 3 mm or asymmetrical. Impairment is indicated if the nonstimulated pupil does not constrict or constricts too slowly. Impairment is indicated if the pupil does not remain constricted. Impairment may be suggestive of cranial nerve 3 or brain stem involvement.

5. *Ocular alignment* is required to achieve visual fusion (the brain's ability to perceive one image). Functionally, a patient will generally complain of diplopia (double vision) if ocular misalignment is present due to a muscle imbalance. Ocular muscle imbalance and weakness is referred to as strabismus.

Note: Administration of this screening can be beneficial within the low-level patient population. If a strabismus is present (a potential for diplopia), identification and normalization of binocular vision may help the patient normalize incoming visual information. When performing this screen (if applicable), ask the patient to fixate on a distant object approximately 25 to 30 inches away.

- In a well-illuminated room, observe both eyes for alignment deviations. Eyes may be skewed in an inward, upward, or horizontal direction compared to the other eye.

- Lower the lighting within the room to allow for use of a penlight.

- Hold the penlight approximately 12 inches from the bridge of the patient's nose.

- Shine the light on the bridge of the patient's nose.

- Compare the position of the corneal reflection in each pupil.

Figure 2.2 shows medial strabismus. The left eye is deviated laterally on forward gaze. Corneal reflection is placed on the medial rim of the pupil. Figure 2.3 shows lateral strabismus. The

Figure 2.2 Medial Strabismus: Accommodation and Convergence

Figure 2.3 Lateral Strabismus

right eye is deviated medially on forward gaze. Corneal reflection is placed on the lateral rim of the pupil. Figure 2.4 shows hypotropia. The left eye is deviated inferiorly on forward gaze. Corneal reflection is placed on superior rim of the pupil. Figure 2.5 shows hypertropia. The left eye is deviated superiorly on forward gaze. Corneal reflection is placed on the inferior rim of the pupil.

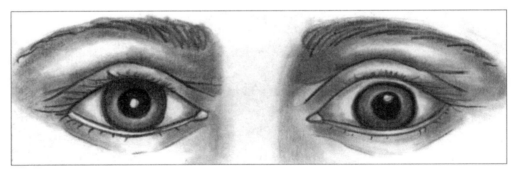

Figure 2.4 Hypotropia

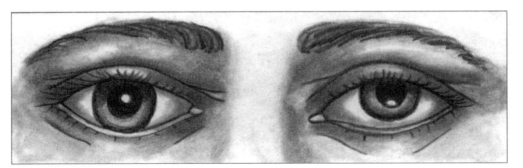

Figure 2.5 Hypertropia

☞ A visual impairment is indicated if one or both eyes are deviated during general observation or the pupillary reflex is not equal in both eyes. Impairment may be suggestive of cranial nerve 3, 4, or 6 involvement, nuclear or supranuclear lesions, or orbital floor fractures leading to entrapment of musculature.

6. *Object fixation* is the ability to locate and focus on a stationary object or target.

- In a well-illuminated room, hold an object or target approximately 16 inches from nose.

- Determine whether the patient can focus and maintain focus on the specified object.

☞ A visual impairment in the area of acuity or binocular function may be indicated if the patient is unable to focus on the target selected. It is essential to establish whether sight is present before administering this component. If sight is not present, fixation will not occur.

Note: Object fixation should be performed only when the patient is able to open the eyes.

7. *Visual pursuit/tracking* is the ability to lock onto and maintain fixation on a moving target across all visual fields.

- In a well-illuminated room with minimal distraction, present the patient with a high-contrast object such as an index card, orange ball, or item of personal relevance to the patient.

Figure 2.6 Visual Tracking (H)

- Hold the object approximately 16 inches from the nose.

- Present the object in mid position to allow for visual fixation.

- Slowly move the object: across the right visual field (see Figure 2.6), toward the left visual field, toward the superior visual field, toward the inferior visual field, and to mid position.

- Slowly move the object diagonally into upper and lower visual fields (see Figure 2.7).

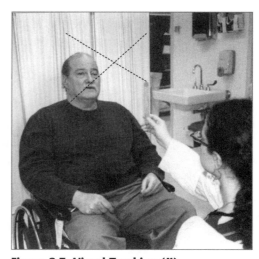

Figure 2.7 Visual Tracking (X)

- Observe the patient's ability to track movement of people within the room.

☛ A visual impairment is indicated if the patient is unable to visually track objects across visual fields. Impairment is indicated if the patient is unable to coordinate the movement of both eyes simultaneously in the same direction and/or is unable to isolate head from eye movement. Ocular movement should be smooth across all visual fields. Impairment may be suggestive of cranial nerve 3, 4, or 6 involvement; visual field cut or visual inattention; cerebellar involvement; fixation; or binocular control deficits.

HIGH-LEVEL PATIENT

A patient is appropriate for a higher-level visual screening when he or she can do the following:

- Follow two- or three-step commands.
- Maintain alertness and sustain attention for more than 15 minutes.
- Communicate with words or an alternative method of communication.
- Tolerate either supported or unsupported sitting.

1. *Functional acuity* is the ability to see detail at both near and far distances. Near distance is that associated with reading distance and performing near tasks such as needlepoint, buttoning, and writing. Far distance is that associated with tasks such as driving, reading signage, discriminating faces and objects, and watching television.

Near distance:

- In a well-illuminated room, place the patient in a sitting position.
- Provide the patient with a near visual acuity chart (Warren Near Text Card or Lighthouse Near Acuity Text Card) held 16 inches away from the eyes (see Figure 2.8).
- Have the patient read top to bottom.
- Continue until the patient misses more than 50% of the line or reading speed is considerably decreased.

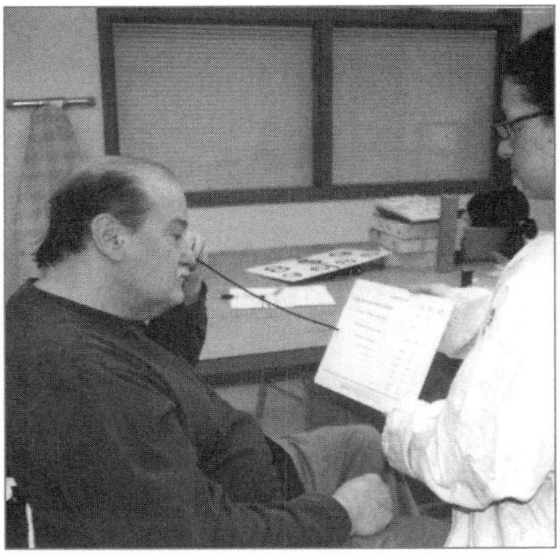

Figure 2.8 Near Vision Acuity Chart

Functional task:

- Ask the patient to read a line of standard-size print in a newspaper or magazine.
- Continue for one page.

☞ A visual impairment is indicated if the patient complains of blurry or fuzzy print, inability to bring the printed material into focus, small print size, or changes the focal length of the reading material. The patient should be able to read at an acuity level of 20/40 or better. Impairments may be suggestive of pathology to the anterior chamber of the eye, optic nerve involvement, macular scotoma, vitreous hemorrhage, occipital lobe involvement, contrast sensitivity, or pupillary control.

Note: Patients should wear their corrective reading glasses during the task. The patient should use binocular vision and not adjust the focal length of the page during the task. If the patient is unable to read letters secondary to aphasia, the Lea Symbols Near Vision Test Card can be used. It is important to have the patient read for a duration of 3 to 5 minutes or the length of one page to observe for fatigue, concentration, and oculomotor control. The therapist should also observe for omission of letters/words, losing place within the text, slow reading speed, and a change in head positioning. If the patient appears confused by the lines or letters, cover all other lines not being read.

Far distance:

- In a well-illuminated room, place the patient in a seated position.

- Provide the patient with a distance acuity chart (Colenbrander or Snellen Acuity Chart; see Figure 2.9).

- Set chart distance 10 or 20 feet from the patient, depending on what chart is being used. If using a Snellen chart, the patient should be seated 20 feet from the target.

Figure 2.9 Distance Acuity Chart

If using a Colenbrander chart, the patient should be seated 1 meter from the target.

- Occlude the patient's left eye with a patch and ask the patient to verbally identify the letters in the specified line.

- Have the patient read top to bottom and left to right.

- Continue until the patient misses more than 50% of the letters on one line.

- Record acuity as the last line on which the patient was successful and note the number of letters missed. For example, 20/40-2 is defined as two out of five numbers missed when reading this particular ototype.

- Repeat with the right eye occluded.

- Repeat with both eyes.

Functional task:

- Have the patient read signs and room numbers either through functional ambulation or from wheelchair level.

☛ A visual impairment is indicated if the patient reads at an acuity level less than 20/40, complains of blurry or fuzzy words, complains of difficulty bringing print/words into focus, or attempts to move closer to the material being read.

Note: The patient should wear corrective distance lenses if applicable. If the patient is unable to identify letters or the numbers on the acuity chart, substitute with the Lea Symbols. The patient should be instructed to maintain head position and focal length/distance throughout the entire test. The therapist should also observe for omission of words, letters, and any signs of changing head position to read. If the patient appears confused by the lines or letters, cover all other lines not being read.

2. *Extraocular range of motion (EOM)* is the degree of range of motion present in each eye within all six cardinal planes. There are six ocular muscles (controlled by cranial nerves 3, 4, and 6) responsible for pulling on the eyeball, thus enabling a person to direct gaze in a particular direction. They are the superior, inferior, medial, and lateral rectus (which have a direct line of pull) and two diagonal muscles, the superior and inferior obliques (which pull the eye in a diagonal plane).

Note: When testing EOM, it is important to move the target in an arclike direction. The patient should be tested monocularly by occluding the nontested eye with a patch. Following a neurological event, it is easier for a patient to coordinate one eye at a time as opposed to two. This test differs from visual tracking in that the therapist focuses on ocular range of motion rather than on the ability to maintain fixation during dynamic movement.

- In a well-illuminated room with minimal distractions, hold a brightly colored target approximately 16 inches from the bridge of the patient's nose.

- Slowly move the object (in an arc) laterally toward the patient's right shoulder and then slowly toward the left shoulder (see Figure 2.6).

- Return the target to mid position.

- Move the target in an upward direction toward the top of the patient's head and then down toward the chin (see Figure 2.6).

- Return to mid position.

- Move the target in an upward diagonal direction toward the upper right temporal area, then in a downward diagonal pattern toward the patient's left shoulder (see Figure 2.7).

- Return to mid position.

- Move the target in an upward diagonal direction toward the upper left temporal area, then in a downward diagonal pattern toward the patient's right shoulder (see Figure 2.7).

- Return to mid position.

- Repeat without occlusion of the eye.

Refer to cranial nerve screening when looking at the breakdown of cranial nerves 3, 4, and 6. The above component addresses the screening of binocular range of motion and quality of movement.

☞ A visual impairment is indicated if an eye is unable to complete the full range of motion within the direction tested. Full range of motion is indicated when the pupil buries a portion of the iris. When testing binocular range of motion, the eyes should move symmetrically and equally in each direction. A visual impairment may suggest cranial nerve 3, 4, or 6 involvement.

3. *Diplopia* (double vision) is a symptom of an ocular muscle imbalance. When light enters the eye, the retina must be stimulated in the exact spot in order for the brain to perceive one image. Functionally, diplopia can present itself horizontally and vertically, affecting all aspects of near and far ADL tasks.

- Place the patient in a seated position.

- In a well-illuminated environment, have the patient reach for a pencil. The pencil should be presented in the superior, inferior, right, left, superior diagonals, and inferior diagonals.

- Ask the patient if diplopia decreases or is absent when closing one eye.

Functional task:

- Have the patient read the clock on the wall at a distance.

- Have the patient read a sentence.

☞ A visual impairment is indicated if the patient complains of a double image/shadowing, overshoots when reaching, behaviorally closes one eye, or alters head position. This may suggest the above cranial nerve involvement, severe refractive errors, or nystagmus.

Note: Diplopia should be assessed if the patient is complaining of double vision, demonstrates overshooting or undershooting, is closing one eye, or has a positive finding during the ocular alignment screening. The patient should wear corrective lenses if applicable. Patients who are confused may not be aware that they are having double vision or able to describe the symptom. Therefore, it is important to observe positioning and behaviors.

4. *Saccades* are quick, precise eye movements that are made during visual scanning or a visual search. Functionally, we use saccades to read and to look for a person in a crowded room. Saccadic eye movements are coordinated at a cerebellar level.

- In a well-illuminated room with minimal distractions, place the patient in a seated position. Sit opposite the patient.

- Hold two brightly colored targets (different colors) 6 to 8 inches apart and approximately 12 to 16 inches from the patient (see Figure 2.10).

- Ask the patient to hold head still and look from one object to the other when told to.

- While the patient is fixating on the stated color, move the other target up or down.

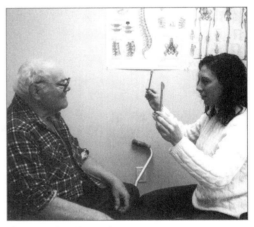

Figure 2.10 Saccades

- Direct the patient to fixate on the newly positioned target.

- Assess midline, superior, inferior, and right and left quadrants.

☞ A visual impairment is indicated if the patient demonstrates undershooting or overshooting between targets, shifting into all fields of gaze, nystagmus, and/or the inability to isolate head and eye movement.

Note: Repeat the task approximately five to seven times until the patient's pattern can be observed.

5. *Accommodation* is a three-step process involving a change in the thickness of the lens, convergence of the eyes, and pupillary constriction. Saccades enable one to adjust to changes in focal length and achieve the sharpest focus. Functionally, accommodation allows us to quickly change focus as we look at our speedometer and the road when driving. Accommodation allows students to copy notes from the board to notebooks without blurring vision.

Screening 1:

- In a well-illuminated room, place the patient in a seated position. Sit opposite the patient.

- Present the patient with a bright target held approximately 20 inches from the eyes (see Figure 2.2).

- Slowly move the target toward the bridge of the nose as the patient is told to fixate on the moving target.

- Measure the distance when the patient states that he or she is no longer able to focus clearly on the target.

Screening 2:

- Have the patient fixate on a distant object approximately 2 feet away.

- Present a bright object within 12 inches from the bridge of the patient's nose.

- Ask the patient to look at the distant object and then the near object.

- Repeat five times.

- Ask the patient whether either the near or the distant object is difficult to bring into focus.

Functional task:

- Does the patient complain of blurred vision during oral hygiene, cosmetic application, and shaving?

☞ A visual impairment is indicated when the eyes fail to converge equally, pupillary constriction is delayed or absent, or the patient is unable to maintain clear acuity within 4 to 5 inches from the bridge of his nose. Functionally, the patient complains of blurred vision, difficulty focusing, and excessive blinking. Impairment may be suggestive of cranial nerve 3, convergence insufficiency, or binocular dysfunction.

Note: The patient should wear corrective lenses if applicable.

6. *Convergence* is the ability of both eyes to move medially during a near task. Convergence is controlled by the brain stem. Functionally, it allows a patient to perform near tasks such as reading.

- In a well-illuminated room with minimal distractions, present a bright target or familiar object to the patient.

- Hold the target approximately 30 inches from the bridge of the nose or at a distance the patient states he or she sees only one target (see Figure 2.2).

- Ask the patient to follow or keep his or her eye on the target.

- Slowly move the object toward the bridge of the nose.

- Observe the quality of ocular movement.

- Ask the patient to state when he or she sees two targets.

- Observe and measure the distance.

☞ A visual impairment is indicated if one or both eyes fail to move medially and symmetrically toward the nose. An impairment is indicated if diplopia is present at a distance greater than 4 inches from the eyes. Visual impairment is suggestive of convergence insufficiency and/or deficits in convergence/divergence.

7. *Visual field* is the space one is able to see when looking straight ahead. The normal field of vision is approximately 160° binocularly. Monocular field of vision is 60° superiorly, 75° inferiorly, 60° nasally, and 100° temporally. Functionally, we depend on our visual field when driving, looking for a particular address, shopping in a supermarket, and selecting clothing.

Confrontation testing:

- In a well-illuminated room, place the patient in a seated position. Sit opposite from the patient at approximately 1 meter.

- Occlude the patient's right eye with an eye patch.

- Instruct the patient to fixate on the therapist's left eye.

- Place both arms behind the patient's head (out of the patient's field of view; see Figure 2.11).

- Slowly bring one arm into the patient's field of view, wiggling only one finger.

- Present the target four times.

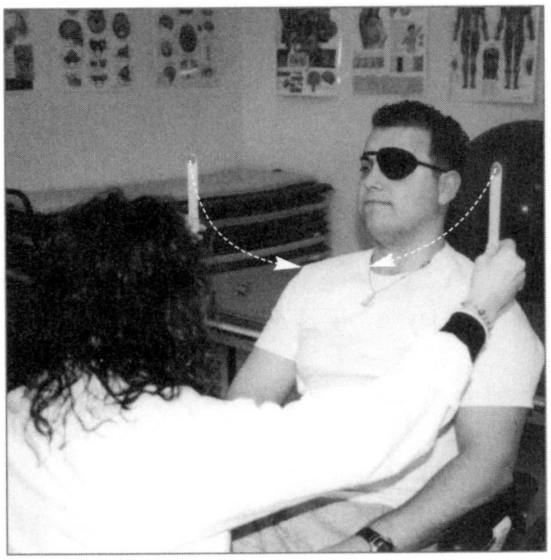

Figure 2.11 Confrontation Testing

- Test the superior, left, and right visual fields, and the inferior visual field.

- Instruct the patient to say "now" or raise his or her hand when he or she first sees the wiggling finger.

- Repeat with the left eye occluded.

Functional observation:

- Observe the patient move through a crowded environment either from wheelchair level or via ambulation.

- Have the patient read a paragraph of standard-size newsprint.

- Have the patient address an envelope.

- Have the patient scan the therapy closet for specific items.

- Observe the patient perform a morning ADL task, such as grooming or dressing.

- Observe body position.

☞ A visual impairment is indicated if the patient is unable to or demonstrates delay in identifying the stimulus in the presented visual field, collides with obstacles within one of the visual fields, demonstrates abbreviated visual scanning in any of the visual fields, omits words or letters when reading, demonstrates skewed/drifted writing, visually searches only half of her environment, presents with postural deviations to the right or left, or complains of feeling off balance. Impairment suggests optic tract or chiasm involvement.

Note: When performing visual field screening, the patient should wear corrective lenses if appropriate. When performing confrontation testing, ensure that the patient is fixating and maintains fixation on the therapist's specified eye when presenting the wiggling stimulus. It is a good idea to administer the stimulus in an unpredictable manner to avoid patient anticipation. When administering the functional-oriented tasks, present reading items in midline and discourage the patient from placing them on the left or right extremes. Behaviorally, it is important to observe how the patient performs the task. Refer to the cranial nerve screening section for interpretation of visual field pathology.

8. *Visual attention* is the ability to visually direct attention within all visual fields (superior, inferior, right, and left) encompassing intra- and extrapersonal space.

Functional observation:

- Observe the patient reading a newspaper or hospital/restaurant menu.

- Ask the patient to write down his demographic information.

- Have the patient select items and perform an oral/facial hygiene task. Men should shave, and women should apply make-up.

- Observe the patient getting dressed.

■ Observe the patient self-propel a wheelchair around the hallway.

■ Observe the patient ambulate around the room, therapy gym, and hospital environment.

■ Observe eating.

■ Observe body position.

☞ A visual impairment is indicated if the patient demonstrates an abbreviated and disorganized visual scan, skews or drifts when writing, attends to only his or her body (intrapersonal space), collides with obstacles during mobility, hesitates to cross midline when an object is placed to one side, or fails to acknowledge or notice persons on the involved side.

Note: Screenings for visual field deficits and visual attention are similar and overlap. It is a good idea when screening for visual attention and visual field deficits to consider the above functional observations. When performing a visual attention screening, observe for impulsive, disorganized behavior.

9. *Contrast sensitivity* is the ability to detect subtle changes in contrast between the background and foreground. Functionally, it is essential to determine varying grades of contrast for tasks such as reading, laundry, driving, cooking, and negotiating curbs and stairs.

■ Ask a patient to fill a clear glass with water from a pitcher to within 1 inch from the rim of the glass.

■ Ask the patient to sort a pile of white laundry.

■ Ask the patient to distinguish blue, black, brown, and purple material.

■ Observe a patient negotiate steps and curbs.

☞ A visual impairment is indicated if the patient has difficulty determining the liquid level in a clear glass, requires additional lighting to perform tasks such as self-care and reading, and has difficulty distinguishing among (a) white, beige, and tan and (b) blue, black, purple, and brown. In addition, the patient demonstrates a hesitation when negotiating curbs or steps or trips. Impairment suggests macular involvement.

10. *Functional mobility* is vital to determine how well a patient is able to navigate in his or her environment. A functional mobility screening incorporates all of the above visual screening methods.

■ Select a hospital or community-based task such as shopping in a market, bank, or taking a bus trip.

■ Consider the following during the task:

– Is the patient scanning all sides of the environment?

– Is he or she able to avoid obstacles?

– Can he or she detect changes in surfaces such as curbs, steps, and slopes?

– Can he or she read signage such as street signs, building numbers, and traffic lights?

– Can he or she use an elevator and escalator?

☞ A visual impairment is indicated if the patient has difficulty with any of the above.

RED FLAGS
SIGNS AND SYMPTOMS OF VISUAL IMPAIRMENT

A red flag is an indicator or symptom of dysfunction. If a patient is observed displaying any of the following red flags, it is likely that the patient possesses visual impairments.

VISUAL ACUITY IMPAIRMENTS: NEAR

- Complains of blurry or fuzzy print.
- Is unable or has difficulty reading print.
- Complains of print being too faint or small.
- Complains of lighting issues.
- Blinks excessively.
- Rubs eyes.
- Changes the focal length of the reading material (moving the material toward or away from self).
- Shifts the page position.
- Views out of the corner of eye.
- Has difficulty performing fine motor tasks such as buttoning or cutting with a scissors.

VISUAL ACUITY IMPAIRMENTS: FAR

- Has difficulty reading signs.
- Has difficulty identifying faces and objects.
- Has difficulty driving.
- Squints and blinks excessively.
- Adjusts reading material closer to eyes.

BINOCULAR VISION IMPAIRMENTS: DIPLOPIA

- Complains of double images, either horizontal or vertical.
- Complains of blurred or shadowed vision.
- Complains of headache or eyestrain/fatigue.

- Has difficulty with depth perception.

- Overshoots or undershoots during a reaching activity.

- Shuts or squints one eye.

- Repositions task or self.

- Tilts head or positions it abnormally.

- Complains of nausea.

BINOCULAR VISION IMPAIRMENTS: CONVERGENCE INSUFFICIENCY

- Complains of losing place when reading or writing.

- Has difficulty performing tasks close up.

- Skips words or lines when reading.

- Complains of intermittent diplopia or blurred vision.

- Squints.

- Complains of headache or eye strain/fatigue during reading.

OCULOMOTOR DEFICITS: OCULAR RANGE OF MOTION

- Has a decrease in ocular range of motion.

- Displays slowness or unequal eye movements.

- Complains of diplopia.

- Complains of difficulty focusing.

- Has disconjugate gaze.

- Complains of headache or eyestrain/fatigue.

- Displays attentional deficits.

- Shows deficits in depth perception.

- Functionally overshoots or undershoots.

OCULOMOTOR DEFICITS: PURSUITS, SACCADES, AND FIXATION

- Has difficulty reading text.

- Skips words and lines.

- Has difficulty with page navigation.

- Displays unnecessary head movements.

- Has difficulty coordinating visually guided movements.

- Is unable to fixate on an object and sustain it.

- Shows jerky eye movements during a reading or tracking task.

- Complains of losing place during reading or a visual searching task.
- Complains of swirling print.
- Decreases speed or increases time required when looking for objects.
- Displays deficits in balance.

ACCOMMODATIVE DISORDERS

- Complains of blurred vision during grooming, buttoning, and shaving.
- Blinks excessively.
- Complains of swirling or moving print.
- Has difficulty reading or performing tasks up close.
- Complains of difficulty focusing.
- Shows difficulty focusing when reading at a distance and writing up close.
- Complains of headaches or eyestrain/fatigue.
- Has sensitivity to light.

VISUAL ATTENTION DEFICITS

- Shows indented or abbreviated visual scanning.
- Has decreased attention to both intra- and extrapersonal space.
- Has decreased attention to visual detail on affected side.
- Skews body position.
- Shows increased obstacle collision on affected side.
- Grooms only on one side of body.
- Dresses or undresses only one side of body.
- Reads only one side of food menu.
- Displays impulsive behavior.
- Rushes through task.
- Shows disorganized visual scanning.
- Is reluctant to recheck work.
- Wears eyeglasses incorrectly (arm not placed behind ear).

VISUAL FIELD DEFICITS

- Misreads literature.
- Is unable to read.
- Has absent or poor eye contact.

- Displays difficulty judging distances.
- Has poor navigational skills/gets lost.
- Complains of seeing only one-half of an image.
- Complains of images/objects darker on one side.
- Shows abbreviated scanning.
- Omits letters or words.
- Scans in disorganized fashion.
- Reluctant to change head position.
- Shows increased muscle tone in the upper and/or lower extremities secondary to postural insecurity.
- Trails upper extremity or cruises during ambulation.
- Is hesitant or unable to ambulate in crowded environments.
- Shows difficulty identifying details of complex information.
- Has deficits in ADL.
- Drifts off line when writing.
- Bumps into objects.
- Holds head to one side.
- Has difficulty gathering and identifying objects for morning routine.
- Shows difficulty with clothing selection in large closet.

IMPAIRED CONTRAST SENSITIVITY

- Has difficulty recognizing faces.
- Has increased falls or tripping.
- Displays difficulty distinguishing among colors: (a) beige, white, and tan and (b) brown, black, blue, and purple.
- Has difficulty cutting nails.
- Has difficulty fastening or tying clothing of similar color.
- Has difficulty pouring liquids.
- Has difficulty cutting or eating when food and plate are of similar color.
- Displays increased collision in poorly illuminated environments.
- Has difficulty with stairs.

AVAILABLE IN-DEPTH ASSESSMENTS

If functional visual screening detects impairment, the following in-depth assessments are available:

Assessment of Motor and Process Skills (AMPS)
A. G. Fisher
Children and adults with developmental, neurologic, and musculoskeletal disorders
AMPS Project
Occupational Therapy Building
Colorado State University
Fort Collins, CO 80523

Behavioral Inattention Test (BIT)
B. Wilson, J. Cockburn, P. Halligan
Adults with visual neglect secondary to neurologic impairment
National Rehabilitative Services
117 North Elm St.
PO Box 1247
Gaylord, MI 49735

Benton Visual Perception Test
A. L. Benton
Children, adolescents, and adults with visual–perceptual processing impairments
Psychological Corporation
555 Academic Ct.
San Antonio, TX 78204-2498

Brain Injury Visual Assessment Battery for Adults (BIVABA)
M. Warren
Adults with visual deficits secondary to neurologic impairment
Precision Vision
745 North Harvard Ave.
Villa Park, IL 60181

Canadian Occupational Performance Measure (COPM)
M. Law, S. Baptiste, A. Carswell, M. A. McColl, H. Polatajko, N. Pollock
Children and adults with cognitive, psychosocial, neurologic, and/or musculoskeletal impairment

Canadian Association of Occupational Therapists
110 Eglinton Ave. West, 3rd Floor
Toronto, ON M4R 1A3 Canada

Central Visual Field Testing for Low-Vision Patients
D. C. Fletcher
Adults with visual deficits
Mattingly International, Inc.
938-K Andreasen Dr.
Escondido, CA 92029

Comprehensive Test of Visual Functioning (CTVF)
S. L. Larson, E. Buethe, G. J. Vitali
Children, adolescents, and adults with neurological impairment
Slosson Educational Publication, Inc.
PO Box 280
East Aurora, NY 14052-0280

Elemental Driving Simulator (EDS) and Driving Assessment System (DAS)
R. Giatnutsos, A. Campbell, A. Beattie, F. Mandriota
Adults with neurologic disorders
Life Sciences Associates
One Fenimore Rd.
Bayport, NY 11795-2115

Erhardt Developmental Vision Assessment (EDVA)
R. Erhardt
Individuals suspected of visual motor deficits
Therapy Skill Builders
Psychological Corporation
PO Box 998
Odessa, FL 33556-9908

Falls Efficacy Scale (EFS)
M. E. Tinetti, D. Richman, L. Powell
Elderly persons with balance deficits
Tinetti, M. E., Richman, D., & Powell, L. (1990). Falls efficacy as a measure of fear of falling. *Journal of Gerontology, 45,* P239–P243.

Functional Reach Test (FRT)
P. W. Duncan, D. K. Weinger, J. Chandler,
S. A. Studenski
Adults with balance deficits secondary to neurologic
damage
Duncan, P. W., Weiner, D. K., Chandler, J., &
Studenski, S. A. (1990). Functional reach: A new
clinical measure of balance. *Journal of Gerontology,*
45, 192–197.

Hooper Visual Organization Test (VOT)
H. E. Hooper
Adolescents and adults
Western Psychological Services
12031 Wilshire Blvd.
Los Angeles, CA 90025-1251

Kohlman Evaluation of Living Skills (KELS)
L. Kohlman-Thomson
Adolescents and adults with psychiatric,
developmental, and/or neurologic impairment
American Occupational Therapy Association
4720 Montgomery Ln.
PO Box 31220
Bethesda, MD 20824-1220

MN Acuity Chart
Adults with visual impairments
Lighthouse Low Vision Products
36-02 Northern Blvd.
Long Island City, NY 11101

Pepper Test (VSRT)
Adolescents and adults with visual impairment
Mattingly International, Inc.
938-K Andreason Dr.
Escondido, CA 92029

Revised Sheridan Gardiner Test of
Visual Acuity
M. D. Sheridan, P. Gardiner
Children ages 5 and over
Keeler Instruments, Inc.
456 Parkway
Broomall, PA 19008

Rivermead Perceptual Assessment Battery
(RPAB)
G. Bhavnani, J. Cockburn, N. Lincoln, S. Whiting
Adults with neurologic impairment
Western Psychological Services
12031 Wilshire Blvd.
Los Angeles, CA 90025-1251

Test of Everyday Attention (TEA)
I. H. Robertson, T. Ward, V. Ridgeway,
I. Nimmo-Smith
Adults with cognitive attentional deficits secondary
to neurologic impairment
National Rehabilitation Services
117 North Elm St.
PO Box 1247
Gaylord, MI 49735

Test of Visual–Motor Skills: Upper-Level
Adolescents and Adults (TVMS: UL)
M. F. Garder
Adolescents and adults with neurologic
impairment
Psychological and Educational Publications, Inc.
1477 Rollins Rd.
Burlingame, CA 94010

Visual Attention Test
D. A. Lister, M. F. Garder
Adults with visual attention impairment secondary
to stroke
Lister, D. A. (1984). Apparatus for assessing
vision after stroke. *Canadian Journal of*
Occupational Therapy, 51, 237–240.

Visual Search and Attention Test (VSAT)
M. R. Trenerry, B. Crosson, J. DeBoe, W. R. Lever
Adults with visual–spatial deficits secondary to
neurologic impairment
Psychological Assessment Resources, Inc.
PO Box 998
Odessa, FL 33556-9908

For further information about the these assessments, refer to the following resources:

Asher, I. E. (1996). *Occupational therapy assessment tools: An annotated index* (2nd ed.). Bethesda, MD: American Occupational Therapy Association.

Impara, J. C., Plake, B. S., & Murphy L. L. (Eds.). (1998). *The thirteenth mental measurements yearbook.* Lincoln: University of Nebraska Press.

Murphy, L. L., Impara, J. C., & Plake, B. S. (Eds.). (1999). *Tests in print: An index to tests, test reviews, and the literature on specific tests* (Vols. 1 & 2). Lincoln: University of Nebraska Press.

Plake, B. S., Impara, J. C., & Murphy, L. L. (Eds.). (1999). *The supplement to* The Thirteenth Mental Measurements Yearbook. Lincoln: University of Nebraska Press.

Referral to an ophthalmologist or behavioral optometrist should be made when screening has detected significant visual impairment.

Reference
Warren, M. (1997, April). *Evaluation and treatment of visual perceptual dysfunction in adult brain injury.* Workshop presented at Warren Visual Perceptual Dysfunction Conference, Baltimore, MD.

FUNCTIONAL VISUAL SCREENING

Patient Name _____   Date _____

Diagnosis _____
_____

VISUAL HISTORY

History of eye trauma	☐ yes	☐ no	
History of eye disease	☐ cataracts	☐ macular degeneration	
	☐ glaucoma	☐ diabetic retinopathy	
Use of glasses	☐ yes	☐ no	
History of falls	☐ yes	☐ no	
Recent changes in vision	☐ yes	☐ no	
Difficulty with ADL or mobility	☐ yes	☐ no	
Visual complaints	☐ diplopia	☐ headache	☐ blurred/fuzzy vision
	☐ eyestrain	☐ difficulty reading	☐ decreased balance

OBSERVATION

Eyelid function	☐ intact	☐ impaired	comment: _____
Does lighting change eye opening/closing?	☐ yes	☐ no	comment: _____
Can patient locate or fixate on a target/object?	☐ yes	☐ no	comment: _____
Can patient track object?	☐ yes	☐ no	comment: _____
Do both eyes move together?	☐ yes	☐ no	comment: _____
Are any ocular deviations noted?	☐ yes	☐ no	comment: _____
Is there a gaze preference?	☐ yes	☐ no	comment: _____
Is there excessive blinking?	☐ yes	☐ no	comment: _____

VISUAL ACUITY   Right (OD)   Left (OS)   Both Eyes (OU)

Optokinetic response (only for coma/low level)	☐ yes	☐ no	comment: _____
Near acuity	_____OD	_____OS	_____OU
Can the patient read standard printed material?	☐ yes	☐ no	comment: _____
Far acuity	_____OD	_____OS	_____OU
Acuity with ambulation	☐ intact	☐ impaired	comment: _____
Dilated	_____OD	_____OS	_____OU
Constricted	_____OD	_____OS	_____OU
Normal	_____OD	_____OS	_____OU

PUPILLARY RESPONSE

	_____OD	_____OS	
	☐ intact	☐ impaired	comment: _____

OCULAR ALIGNMENT

	_____OD	_____OS	
Reflection equal	☐ intact	☐ impaired	comment: _____
Reflection deviated superiorly	☐ intact	☐ impaired	comment: _____
Relation deviated inferiorly	☐ intact	☐ impaired	comment: _____
Relation deviated medially	☐ intact	☐ impaired	comment: _____
Relation deviated laterally	☐ intact	☐ impaired	comment: _____

OBJECT FIXATION

Can the patient fixate on a target?	☐ yes	☐ no	comment: _____
Do the eyes appear aligned?	☐ yes	☐ no	comment: _____

VISUAL PURSUIT

Can the patient track object/person?	☐ yes	☐ no	comment: _____
Are there any restrictions across the visual fields?	☐ yes	☐ no	comment: _____
Is nystagmus noted?	☐ yes	☐ no	comment: _____
Are eyes moving together?	☐ yes	☐ no	comment: _____

DIPLOPIA

	☐ yes	☐ no	
	☐ horizontal	☐ vertical	
	☐ intermittent	☐ constant	
	☐ near	☐ far	
	_____OD	_____OS	
Is there observed over- or undershooting?	☐ yes	☐ no	comment: _____
	☐ intact	☐ impaired	

ACCOMMODATION

Screening 1	☐ intact	☐ impaired
Screening 2	☐ intact	☐ impaired
If impaired check one:	☐ near target	☐ distant target

CONVERGENCE

	☐ intact	☐ impaired	_____distance at which diplopia is present

VISUAL FIELD

Screening 1: Confrontation	☐ intact	☐ impaired	explain: _____
Screening 2: Functional			
Navigation in a crowded environment	☐ intact	☐ impaired	explain: _____
Reading text	☐ intact	☐ impaired	explain: _____
Unstructured writing	☐ intact	☐ impaired	explain: _____
Unstructured scanning of extrapersonal space	☐ intact	☐ impaired	explain: _____
ADL performance	☐ intact	☐ impaired	explain: _____
Body position	☐ turned R	☐ turned L	☐ neutral

VISUAL ATTENTION

Reading newspaper	☐ intact	☐ impaired	explain: _____
Unstructured writing	☐ intact	☐ impaired	explain: _____
Grooming task	☐ intact	☐ impaired	explain: _____
Dressing task	☐ intact	☐ impaired	explain: _____
Functional mobility	☐ intact	☐ impaired	explain: _____
Eating	☐ intact	☐ impaired	explain: _____
Body position	☐ intact	☐ impaired	explain: _____

CONTRAST SENSITIVITY

Fill glass	☐ intact	☐ impaired
Sort wash	☐ intact	☐ impaired
Color discrimination	☐ intact	☐ impaired
Negotiation of steps/curbs	☐ intact	☐ impaired
Is the patient having difficulty identifying faces?	☐ intact	☐ impaired

VISUAL SKILLS DURING AMBULATION

Scanning of the environment	☐ intact	☐ impaired
Is the patient having collisions?	☐ intact	☐ impaired
Ability to detect curbs/steps	☐ intact	☐ impaired
Ability to read signs and landmarks	☐ intact	☐ impaired
Ability to use escalator/elevator	☐ intact	☐ impaired

Perceptual
SCREENING

FUNCTIONAL IMPLICATIONS OF PERCEPTUAL IMPAIRMENT

Perception is the ability to interpret or attach meaning to sensory information from the external and internal environments. Perceptual impairment more often involves *dysfunction of the right hemisphere* rather than the left, although some perceptual disorders result from left hemisphere damage (e.g., some perceptual language disorders). Right hemisphere perceptual disorders involve a distortion of the physical environment. The patient's perception of the environment and his or her body become distorted. Several classifications of perception exist:

- Visual perception

- Visual–spatial perception

- Tactile perception

- Body schema perception

- Language perception

- Motor perception.

Patients with *visual perception disorders* have difficulty identifying and recognizing familiar objects and people (despite having intact visual anatomical structures). People and objects in their environment appear to have distorted physical properties. For example, patients may be unable to recognize family members or even their own face as observed in a mirror. They may be unable to interpret the meaning of traffic light colors and thus present a safety risk to themselves and others. Objects may appear bigger, smaller, heavier, or lighter than they really are. A bag of groceries may appear smaller and lighter than it actually is, causing the patient to drop the bag and spill its contents as he or she attempts to lift the bag.

Patients with *visual–spatial perception disorders* have difficulty accurately interpreting the spatial relationships between their bodies and objects in the environment. Cars may appear farther away than they actually are, causing the patient to become a safety risk when attempting to cross the street or drive. Drinking glasses and cups may appear closer than in actuality, thus causing the patient to frequently spill liquids. Objects in the foreground may blend with the background, causing the patient difficulty when attempting to retrieve specific items. For example, freshly laundered white shirts placed on a bed with white sheets may become invisible to the patient. A bar of white soap may become indistinguishable from the background of the white sink. The patient may also commonly lose a sense of direction when traveling in the community and become lost, even when traveling familiar routes. Similarly, when using concepts of direction, right and left often become reversed. Patients who are following directions to a specific destination may commonly confuse the concepts of right and left and easily become lost.

Patients with *tactile perception disorders* have difficulty attaching meaning to objects in the environment by touch alone. Patients commonly have difficulty reaching into their pockets for specific coins because they are unable to discriminate between coins by touch alone. Similarly, reaching into pockets or a dark-colored handbag for keys, pens, or credit cards is equally challenging. Computer touch-typing would be very difficult. Patients would likely be unable to distinguish one key from the others without visual cues.

Patients with *body schema perception disorders* have a distorted awareness of their body (or specific parts of their body, such as a hemiplegic limb). They may not recognize their limb (e.g., a hemiplegic upper extremity) as their own. They may dissociate from their hemiplegic limb to such an extent that they ask for the limb to be taken out of their bed or taken off of their wheelchair lap tray. Patients with left neglect commonly fail to dress the left half of their body, neglect to shave the left side of their face, or put make-up on one side of their face while neglecting the opposite side. In addition to an unawareness of one side of their body, patients may be unaware that the entire left side of the environment exits. Such patients will frequently wheel themselves into walls and neglect to see the food on the left side of their plate. They commonly fail to see the entire left side of a written page when reading and instead begin to read each sentence at its midpoint.

Patients with body schema disorders often become safety risks. They may be unaware that their hemiplegic limb is dragging on the floor when they are using a wheelchair or that their hemiplegic arm is getting caught in the spokes of the wheelchair as it dangles over the wheelchair arm. They may injure themselves or others by wheeling themselves into walls or furniture on the left side of the environment. During meal preparation, they may burn their involved limbs without an awareness of having sustained tissue damage. Similarly, scalding from hot water may occur to their involved side without their awareness.

Patients with *language perception disorders* have difficulty with the expression and/or the comprehension of language. Patients with language disorders resulting from *left hemisphere damage* will have difficulty understanding the abstract or symbolic meaning of language. Instead, all language will be interpreted literally; thinking will be concrete. Although such patients may accurately interpret the literal meaning of a conversation, they will fail to understand the meaning conveyed through each person's vocal tone. The emotional tone of the conversation—whether someone is angry, surprised, or joking—will be lost. Conversely, patients whose language disorder results from *right hemisphere damage* will be able to understand the unspoken tone of a conversation but may fail to understand the literal meaning of someone's words. Theorists have made analogies to pets who cannot understand the literal meaning of every human word but can nevertheless understand the human's meaning by vocal tone alone.

Patients with *receptive language disorders* will have difficulty understanding what others are saying. Because they do not understand others, they will have difficulty communicating. Patients with *expressive language disorders* can understand what others have said but are unable to communicate their own needs in an understandable way. Such patients are often highly frustrated because they are unaware that their own attempts at communication are unclear and may even sound like gibberish. Patients with receptive language disorders may also exhibit *dyslexia,* difficulty reading, and *asymbolia,* difficulty comprehending gestures, whereas patients with expressive language disorders may exhibit *agrommation* (difficulty arranging words sequentially to form intelligible sentences), *agraphia* (difficulty writing intelligible sentences), and *acalculia* (difficulty performing mathematical calculations).

Patients with *motor perception disorders* have difficulty with motor planning, that is, the ability to implement a specific motor action pattern, as required by environmental demands. Patients with motor perception disorders have a distorted perception of the motor strategies required to negotiate their environment. Patients may have difficulty understanding the motor demands of a task. For example, a patient may not understand that a shirt is an article of clothing to be worn on the torso and upper extremities. Patients may understand the demands of the motor task but may have lost the specific motor plan for that specific task. For example, the patient recognizes that a shirt is an article of clothing to be worn on the torso and upper extremities but may be unable to access the specific motor plan for donning a shirt. He or she may instead access an inappropriate motor plan and attempt to place the shirt on the legs.

PERCEPTUAL SCREENING PROCEDURES

VISUAL PERCEPTION DISORDERS

1. *Visual agnosia* is an umbrella term for the inability to identify and recognize familiar objects and people (despite intact visual anatomical structures). Lesions that cause visual agnosias often are located in the right hemisphere occipital lobe or posterior multimodal association area.

- Show the patient several familiar items, one at a time (e.g., pencil, eyeglasses, hairbrush, keys, wristwatch).

- Ask the patient to identify the object.

It is important to discriminate between visual agnosia and aphasia. If the patient is experiencing word-finding difficulties, offer the patient a choice of three answers. Ask the patient to indicate the correct choice by shaking his or her head "yes." Visual agnosia may be indicated if the patient is unable to name four out of five items.

2. *Prosopagnosia* is the inability to identify familiar faces because the patient cannot perceive the unique expressions of facial muscles that make each human face different from another. Prosopagnosia often results from right hemisphere damage.

- Show the patient photographs of familiar people (e.g., world leaders, celebrities, sports figures). Make sure that the famous people shown are individuals whom the patient would recognize, that is, people from the patient's own era and culture.

■ Show the patient photographs of familiar family members.

■ Ask the patient to identify the names of the people in the photographs. If the patient is unable to think of the individuals' names, ask the patient what the famous people are known for. Or offer the patient a choice of three answers. If the patient is an expressive aphasic, allow him or her to nod "yes" in response to the correct answer.

■ Hold a mirror in front of the patient's face. Ask the patient to identify the person reflected in the mirror. Or give the patient a photo of him- or herself and ask the patient to identify the person shown.

☞ Prosopagnosia may be indicated if the patient is unable to identify familiar family members, famous people, and/or him- or herself as reflected in a mirror or photograph.

3. *Simultanognosia* is the inability to interpret a visual stimulus as a whole. It often results from right hemisphere damage.

■ Show the patient photographs of detailed scenes (e.g., a farm with animals or crops, a zoo, a city street, the inside of a grocery store).

■ Ask the patient to describe the scene in detail.

■ Or, take the patient to a window and ask him or her to describe the scene outside the window.

Patients with simultanognosia may be able to pick up small details from the scene but are unable to integrate the details in order to understand the scene as a whole. For example, when looking at a photograph, they may see a car, a person walking, and a storefront window, but they may be unable to understand that they are looking at photo of a busy city street. Often, patients with simultanognosia confabulate in an effort to fill in the parts of the scene that they cannot interpret.

☞ Simultanognosia is indicated if the patient is unable to understand the scene as a whole or attempts to fill in the parts he or she cannot interpret through confabulation.

4. *Metamorphopsia* involves a visual distortion of the physical properties of objects so that objects appear bigger, smaller, heavier, or lighter than they really are.

■ Present the patient with a puzzle of simple shapes with differing sizes. Ask the patient to put the correctly sized shape into its corresponding cut-out space.

■ Determine whether the patient has difficulty matching the correctly sized shape to its corresponding cut-out space.

■ Present the patient with several objects (e.g., a mug, a full grocery bag, a thick hardcover book, a bag of cotton balls). Ask the patient to estimate each object's size and weight by observation alone.

■ Present the patient with several different-sized drinking glasses, each filled with water. Ask the patient to put the glasses in order from holding the most water to holding the least water.

☛ Metamorphopsia may be indicated if the patient is unable to accurately place the puzzle pieces into their matching cut-out spaces, if the patient is unable to accurately estimate the size and weight of the presented objects, and if the patient is unable to accurately order the drinking glasses.

5. *Color agnosia* is an inability to remember the appropriate colors for specific objects. Patients with color agnosia appear to forget the color of common objects. For example, a patient with color agnosia may believe that a banana is blue.

■ Ask the patient to name the correct color for the following named objects (do not present the actual object but rather only state the object's name): an apple, an ocean, a fire engine, a stick of butter, coffee. Determine whether the patient is able to accurately identify the color of the named objects.

■ Present the patient with several objects, two of which are inaccurately colored. For example, present the patient with several pieces of plastic fruit, two of which are the wrong color (e.g., paint the banana blue and the grapes pink). Ask the patient to identify any objects that are incorrectly colored.

☛ Color agnosia is indicated if the patient is unable to identify the correct color for named objects and is unable to identify which shown objects are incorrectly colored.

6. *Color anomia* is an inability to remember the names of colors. Color anomia differs from color agnosia in that patients with color anomia may forget the names for colors but would still recognize that a banana is not blue.

■ Show the patient flash cards, each having one distinct, simple color (i.e., red, blue, yellow, green, orange, purple). Ask the patient to name the color.

■ For patients with expressive aphasia, offer the patient three choices for each card.

■ Tell the patient to indicate the correct answer by nodding "yes."

☛ Color anomia is indicated if the patient is unable to accurately identify three out of the six colors shown on each separate flash card.

VISUAL–SPATIAL PERCEPTION DISORDERS

1. *Right–left discrimination dysfunction* is an inability to accurately use the concepts of right and left.

■ Ask the patient to point to his or her own right and left body parts. For example,

– Point to your left elbow.

– Point to your right knee.

– Point to your right shoulder.

– Point to your left foot.

■ Ask the patient to follow your right–left commands as you give him or her right–left directions to walk around the treatment environment. For example,

– Go to the back of the room and open the left door of the closet.

– Go to the desk and open the top drawer on the right.

– Walk down the hall and open the first door on your right.

– Open the window at the far left of the room.

☞ Right–left discrimination impairment is indicated if the patient confuses right and left three out of four times when pointing to right and left body parts. It is also indicated if the patient confuses right and left directions three out of four times when using right–left commands to walk around the treatment environment.

2. *Figure–ground discrimination dysfunction* is an inability to distinguish objects in the foreground from objects in the background.

■ Ask the patient to pick out forks from a kitchen drawer (or table tray) with disorganized, multiple utensils (all having a silver color).

■ Ask the patient to find the white toothbrush, bar of white soap, and white washcloth as they sit against the background of a white bathroom sink.

☞ Figure–ground discrimination impairment is indicated if the patient is unable to pick out more than half of the forks from the other utensils and is unable to easily find the white toothbrush, soap, and washcloth against the white background of the bathroom sink.

3. *Form–constancy dysfunction* is an inability to recognize subtle variations in form or changes in form such as a size variation of the same object.

■ Determine whether the patient is able to identify a familiar object when turned on its side or placed upside down. Ask the patient to identify the following by observation alone (do not allow the patient to manipulate the objects with his or her hands): a plate turned upside down, a fork placed on its side, a pair of scissors opened fully, a doorknob placed on the table in front of the patient.

■ Determine whether the patient is able to categorize a group of shapes having varying sizes. Give the patient several different sizes of triangles, squares, and rectangles (all having the same color).

■ Instruct the patient to sort the shapes into categories. Can the patient sort all of the triangles, squares, and rectangles into separate groups even if the shapes are of varying sizes?

☞ Form–constancy dysfunction is indicated if the patient is unable to identify these familiar objects when they are placed in odd positions. It is also indicated if the patient makes several errors when attempting to sort shapes of varying sizes.

4. *Position in space dysfunction* involves difficulty using concepts relating to positions, such as *up/down, in/out, behind/in front of,* and *before/after.*

- Instruct the patient to follow at least three of the following directions using the above terms:
    - Place the pencil on top of the box. Place the box inside the drawer.
    - Take the pot from underneath the sink and place it on the table. Put the bag of rice inside the pot.
    - Place the comb in front of the [small free-standing] mirror and place the hairbrush behind the mirror.
    - Put the bar of soap on top of the sink and place the hairbrush inside the medicine cabinet.
    - Place the soup can on the top shelf of the cabinet and put the cereal box on the bottom shelf.

☞ Position in space dysfunction is indicated if the patient demonstrates difficulty carrying out two out of three of the these commands.

5. *Topographical disorientation* involves difficulty comprehending the relationship of one location to another. Ask the patient to find his or her way around the treatment facility using verbal directions or a written or pictorial map (provided by the therapist).

- Ask the patient to find his or her way back from the occupational therapy treatment setting to his or her hospital room (or group home room).

- Ask the patient to find his or her way back from the occupational therapy treatment setting to the dining room or cafeteria, lobby, gift shop, and so forth.

- If the treatment setting is large and confusing, ask the patient to find his or her way around only one hospital floor or unit. For example, ask him or her to find the way from his or her hospital room to the nurses' station, to the elevators, and back.

☞ Topographical disorientation is indicated if the patient commonly becomes lost while trying to navigate around the treatment setting using verbal directions and/or a written or pictorial map.

6. *Depth perception dysfunction* (stereopsis) is an inability to determine whether objects in the environment are near or far in relation to each other and in relation to the patient.

- Place several blocks of differing sizes on a table in front of the patient. Ask the patient to identify which block is the farthest away, which block is closest, and which blocks are at mid-range between the farthest and the closest blocks.

■ Place the patient in front of a window. Ask him or her to identify which objects are closest to the building, which are farthest, and which are at mid-range.

☛ Depth perception impairment is indicated if the patient is unable to accurately determine which blocks are closest and farthest and at mid-range. It is also indicated if the patient is unable to accurately determine which objects (outside the window) are closest to the building, which are farthest, and which are at mid-range.

TACTILE PERCEPTION DISORDERS

Tactile agnosia is the umbrella term for the inability to attach meaning to somatosensory data, or tactile data. Tactile agnosia commonly results from lesions to the secondary somatosensory area. One's touch and pain and temperature receptor anatomy remain intact. The ability to attach meaning to somatosensory data is referred to as *cortical sensation.*

1. *Astereognosis* is the inability to identify objects by touch alone.

■ Instruct the patient to identify the following by touch alone (occlude the patient's vision): a key, a quarter or other coin, a paper clip, a wristwatch, a pencil.

■ Test astereognosis in one hand at a time.

■ Alternate the sequence of presented objects for each hand.

☛ Astereognosis is indicated if the patient is unable to identify three out of the five objects presented. If the patient is able to identify the objects using one hand but not the other, the lesion is indicated in the contralateral hemisphere.

2. *Ahylognosia* is the inability to discriminate between different types of materials by touch alone.

■ Instruct the patient to identify the following materials by touch alone (occlude the patient's vision): a cotton ball; a piece of metal, such as a washer or large nail; a piece of cloth, such as velvet; a piece of rubber, such as a rubber band or theraband; a piece of wood, such as a wooden block.

■ Test ahylognosia in one hand at a time.

■ Alternate the sequence of presented materials for each hand.

☛ Ahylognosia is indicated if the patient is unable to identify three out of the five presented materials. If the patient is able to identify the materials using one hand but not the other, the lesion is indicated in the contralateral hemisphere.

3. *Amorphagnosia* is the inability to discriminate between different forms by touch alone.

■ Instruct the patient to identify the following forms by touch alone (occlude the patient's vision): a triangle (a triangular styrofoam form), a square (a square block), a circle (a ball or circular styrofoam form), a rectangle (a rectangular block), a star (a styrofoam star).

■ Test amorphagnosia in one hand at a time.

■ Alternate the sequence of presented forms for each hand.

☞ Amorphagnosia is indicated if the patient is unable to identify three out of the five presented forms. If the patient is able to identify the forms using one hand but not the other, the lesion is indicated in the contralateral hemisphere.

4. *Two-point discrimination* is the inability to determine whether one has been touched by one or two points.

■ Use an esthesiometer to determine whether the patient can identify whether he or she has been touched by one or two points.

■ Instruct the patient that you are going to touch his or her fingertip with one or two points—test two-point discrimination on one hand at a time. Ask the patient to state whether he or she has been touched by one or two points (occlude the patient's vision). Gradually move the points of the esthesiometer closer together until the patient states that he or she is being touched by one point only. Note the distance between points at this time.

■ Repeat with the opposite hand.

☞ Most people can determine that they have been touched by two points at a minimal distance of 3 mm on the finger tip, 4 mm to 5 mm on the middle phalanx, 6 mm to 7 mm on the proximal phalanx, and 7 mm to 10 mm on the palm. If the patient has normal two-point discrimination in one hand but not the other, the lesion is indicated in the contralateral hemisphere.

5. *Agraphesthesia* is the inability to interpret letters written on the palmar surface of one's hand.

■ Instruct the patient that you are going to write letters on his or her hand with your fingertip (occlude the patient's vision).

■ Write five letters on the patient's hand, one at a time (e.g., *o, w, t, x, s*). Ask the patient to identify each letter. Test one hand at a time.

☞ Agraphesthesia is indicated if the patient is unable to identify three out of the five letters. If the patient has normal agraphesthesia on one hand but not the other, the lesion is indicated in the contralateral hemisphere.

6. *Extinction of simultaneous stimulation* is the inability to determine that one has been touched on both the involved and uninvolved sides; the neural sensation of the uninvolved side overrides the ability to perceive touch on the involved side.

■ Touch the patient on two body regions (with the patient's eyes closed):

  – On the involved side

  – On the uninvolved side.

■ Ask the patient to tell you where he or she has been touched (see Figure 3.1).

☞ Extinction occurs when the patient cannot feel the tactile sensation on the involved side because the neural perception of tactile sensation on the uninvolved side overrides the neural perception of the tactile sensation on the involved side.

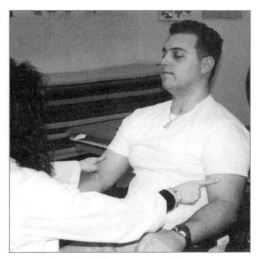

Figure 3.1 Extinction of Simultaneous Stimulation

BODY SCHEMA PERCEPTION DISORDERS

Body schema is the awareness of spatial characteristics of one's own body in space. It is derived from the synthesis of tactile, proprioceptive, and pressure sensory perceptions. Body schema perceptual disorders are more likely to result from right hemisphere lesions in the posterior multimodal association area.

1. *Finger agnosia* is an impaired perception of the relationship of the fingers to each other. Patients with finger agnosia have difficulty identifying and localizing their own fingers.

■ Instruct the patient to carry out the following commands:

  – Show me your thumbs. Show me your index fingers. Show me your ring fingers.

  – Touch your ring finger to your thumb (on the patient's involved side or bilaterally).

  – Touch your index fingers together.

  – Touch your little finger to your thumb (on the patient's involved side or bilaterally).

☞ Finger agnosia is indicated if the patient is unable to demonstrate half of the above commands.

2. *Unilateral neglect* is the inability to integrate and use perceptions from one side of the body and/or one side of the environment. The awareness of the left side of the environment is lost temporarily. Patients more often experience left neglect syndromes rather than right neglect syndromes, although right neglect can occur as well. A left neglect often results from a lesion to the right hemisphere's posterior multimodal association area. Patients with unilateral neglect can be trained to heighten their awareness of the left side of their bodies and the environment.

- Ask the patient to perform the following tasks:

  – Draw a clock. Note whether the patient neglected to fill in the numbers on the left side of the clock (see Figure 3.2).

  – Draw a human figure. Note whether the patient drew only one-half of the body or left off limbs on one side of the body (see Figure 3.3).

- Ask the patient to read a paragraph from a book or magazine page. Note whether the patient begins each sentence at the middle of the page and ignores the words to the left of midline.

- Give the patient a page on which columns of letters appear. Instruct the patient to cross out all of the *H*s. Note whether the patient crossed out only the *H*s on the right half of the page and ignored those on the left half of the page (see Figure 3.4).

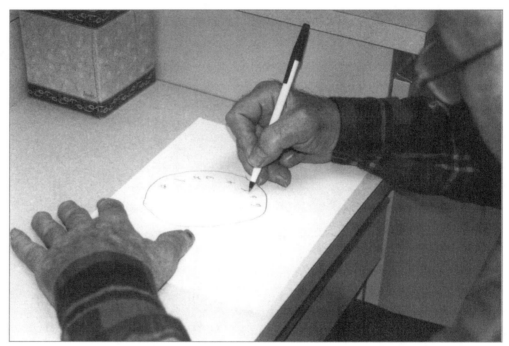

Figure 3.2 Left Neglect: Draw a Clock

Figure 3.3 Left Neglect: Draw a Human Figure

Figure 3.4 Left Neglect: Cross Out *H*s

- Observe the patient while eating. Note whether the patient attends to both sides of the plate or ignores the food on the left half of the plate and meal tray (see Figure 3.5).

- Observe the patient during the morning grooming routine. Note whether the patient fails to shave one side of his face or bathe one half of his or her body.

- Observe the patient while dressing. Note whether the patient is able to dress both halves of his or her body or whether he or she dresses only one side (see Figure 3.6).

Figure 3.5 Left Neglect: Food on Plate

3. *Anosognosia* is an extensive neglect syndrome involving failure to recognize one's paralyzed limbs as one's own. It results from lesions of the right hemisphere. Right hemisphere disorders involve a distortion of the physical environment and one's own body. Although patients with unilateral neglect can be taught to enhance their awareness of the left (or sometimes right) side of their body and environment, patients with anosognosia cannot be taught in the same way. Anosognosia is accompanied by a strange affective dissociation. Patients show extraordinary indifference to their affected limb(s), often asking others to help them remove their affected limb(s) from their wheelchair lap tray or bed as though those limbs were not part of their body. Anosognosia often is a transient state of the patient with acute cerebrovascular accident. Anosognosia will usually resolve as the patient recovers. To determine whether a patient has anosognosia, ask the patient to perform three of the following tasks:

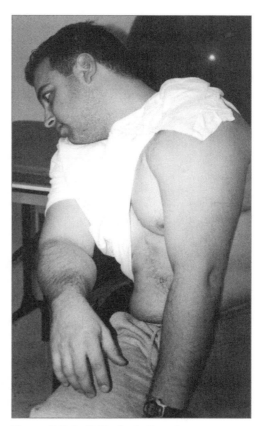

Figure 3.6 Left Neglect: Dressing

- Ask the patient to show you his or her affected upper extremity (ask him or her to show you the left or right arm). If the patient cannot or if the patient can show you only his or her unaffected upper extremity, anosognosia is indicated.

- Take the patient's affected upper extremity and shake hands with that limb. Ask the patient whose hand you are shaking. If the patient cannot recognize that you are shaking hands with his or her affected upper extremity, anosognosia is indicated.

■ Ask the patient to tap the affected leg with his or her unaffected hand (i.e., "tap your left leg with your right hand"). If the patient cannot tap his or her affected leg or can tap only his or her unaffected leg, anosognosia is indicated.

■ Show the patient a colored block. Tell the patient that you are going to put the block somewhere on her body. Place the block on the patient's unaffected thigh.

■ Ask the patient to find the colored block. Then repeat the procedure with the affected lower extremity.

☛ If the patient cannot find the block placed on the affected lower extremity, anosognosia is indicated.

4. *Extinction of simultaneous stimulation* is a form of tactile perception, but it is also considered to be a form of *attentional neglect*. Screening procedures are the same as those listed for tactile perceptual disorders.

■ Touch the patient on two body regions (with the patient's eyes closed):

– On the involved side

– On the uninvolved side.

■ Ask the patient to tell you where he or she has been touched (see Figure 3.1).

☛ Extinction occurs when the patient cannot feel the tactile sensation on the involved side because the neural perception of tactile sensation on the uninvolved side overrides the neural perception of the tactile sensation on the involved side.

LANGUAGE PERCEPTION DISORDERS

Language perception disorders are collectively referred to as the *aphasias.* Aphasia means an impairment in the expression and/or the comprehension of language. There are two primary classifications of aphasia: receptive aphasia and expressive aphasia. *Receptive aphasia* is the impairment in the comprehension of language.

1. *Wernicke's aphasia* is difficulty comprehending the literal interpretation of language. Wernicke's aphasia always results from a left hemisphere lesion in the brain region referred to as *Wernicke's area.*

■ Determine whether the patient can understand the spoken word.

■ Ask the patient simple questions, for example,

– What is your name?

– When is your birthday?

– How old are you?

– Where do you live?

☞ A patient with Wernicke's aphasia will not be able to understand the questions and, thus, will offer answers that do not make sense. However, the patient's verbal responses (while not appropriately answering the questions) will nevertheless be fluent; in other words, the patient's sentences are clear rather than meaningless gibberish as may been seen in expressive aphasia.

2. *Alexia* is the inability to comprehend the written word or the inability to read. Alexia can occur as a result of lesions to either hemisphere; more often the left hemisphere is lesioned.

- Ask the patient to read a simple paragraph (with large print). If the patient uses glasses to read, make sure that he or she is screened with glasses on. It is also important to be certain that the patient is not illiterate and does not have visual field cuts or visual–perceptual problems that would interfere with reading.

☞ An inability to read (in the absence of visual problems, visual–perceptual disorders, and illiteracy) may indicate alexia.

3. *Dyslexia* is the impaired ability to read. Dyslexia is a language problem in which the ability to break down words into their most basic units—phonemes—is impaired. The patient may perceive letters as reversed or sequentially mixed up. Some words in a sentence may be overlooked or left out. If the patient uses glasses to read, make sure that he or she is screened with glasses on. It is also important to be certain that the patient is not illiterate and does not have visual field cuts or visual–perceptual problems that would interfere with reading.

- Rule out a possible previous history of dyslexia before screening.

- Ask the patient to read a simple paragraph, for example,
  The barnyard has five different kinds of animals. There are pigs, goats, cows, sheep, and chickens. A barn cat keeps the mice away. A herding dog keeps the sheep in the pasture. At noon, the farmer's wife milks the cows. At dinnertime, the family sits down together for an evening meal.

☞ Dyslexia is indicated if the patient has difficulty reading the paragraph because of an inability to break down words into phonemes. In other words, the patient will reverse letters or mix up the sequence of letters within words. The above paragraph may appear as follows to a patient with dyslexia:

- The dranyrad five ibffeernt kisnd of namials. hTeer pgis, gatos, cows, shepe, adn chciknes. A darn cat keeds teh ncie waay. A hrebing bog keeps sheeg in the pastreu. At noon, the arfmre's iwfe mills the cwos. At binnretmie te fanliy bon tothgeer rfo an evneing meal.

4. *Asymbolia* is difficulty comprehending gestures and symbols. It is often caused by left hemisphere lesions; however, it can result from right hemisphere lesions as well.

■ Demonstrate the following gestures and symbols to the patient and ask him or her to tell you what they mean. If the patient has word-finding difficulties, offer the patient a choice of three verbal, written, or pictorial options.

– Wave hello/goodbye.

– Shake your head no/yes.

– Point to your watch and gesture "What time is it?"

– Shiver to demonstrate that you are cold; chatter your teeth.

– Gesture with your hand for the patient to come forward.

☞ Asymbolia is indicated if the patient is unable to understand three or more of the these five gestures.

5. *Aprosodia* is impaired comprehension of tonal inflections used in conversation. Patients have difficulty perceiving the emotional tone of someone's conversation. Aprosodia often results from a right hemisphere lesion. Patients can often understand the literal meaning of words but cannot interpret the words' emotional tonal inflections.

■ To determine whether a patient is aprosodic, state a neutral sentence to the patient but change your tonal inflection to sound angry, sad, happy, or disgusted.

■ Ask the patient to identify your emotion. If the patient has word-finding difficulties, offer him or her a choice of three verbal, written, or pictorial answers.

– Say in an angry tone, "That cat came over the fence again last night." Ask the patient to identify your emotion.

– Say in a happy, enthused tone, "That cat came over the fence again last night." Ask the patient to identify your emotion.

– Say in a sad tone, "That cat came over the fence again last night." Ask the patient to identify your emotion.

☞ Aprosodia is indicated if the patient is unable to identify the appropriate emotional tonal inflections in two out of the three sentences.

EXPRESSIVE APHASIA

Expressive aphasia is the inability to express language in a clear, fluent, meaningful way.

1. *Broca's aphasia* is an expressive language disorder in which patients can understand what is spoken to them but cannot express their ideas in an understandable way. Broca's aphasia always results from a left hemisphere lesion in the brain region referred to as *Broca's area*. Often, patients speak in *nonfluent sentences* that do not make sense. Patients may even speak gibberish. Sentences tend to be *telegraphic,* that is, preserving only the essential content words: "Vacation . . . New York." Often, patients will demonstrate semantic or paraphasia sentences in which they substitute a related or unrelated word for the desired target word: *Mother* is used for *wife; dinner* is used for *breakfast.* Because patients can comprehend what is being asked of them, they can sometimes write their answers or point to a desired object or picture of a desired need (e.g., bathroom).

■ Ask the patient to answer the following simple questions:

– What is your name?

– When is your birthday?

– How old are you?

– Where do you live?

☞ If the patient is unable to answer the questions using clear, fluent speech that makes sense, Broca's aphasia is likely indicated.

Note: To determine that the patient understood the question (and that comprehension is intact), allow the patient to attempt to answer the questions in writing; by shaking his or her head "no"/"yes"; or by choosing one of three verbal, written, or pictorial answers provided by the therapist.

2. *Anomia* is the inability to remember and express the names of people and objects. The patient may know the person but cannot remember his or her name. Anomia differs from prosopagnosia: In *prosopagnosia* individuals do not recognize familiar faces. Lesions resulting in anomia can occur in either hemisphere.

■ Present the patient with several familiar objects (e.g., a hairbrush, wristwatch, apple, keys, eyeglasses, pencil). Ask the patient to name the object. If the patient cannot accurately name the object as a result of anomia, allow the patient to correctly name the object by choosing one of three written or verbal answers provided by the therapist (the patient can shake his or her head "no"/"yes" in response to the choices). If the patient can accurately choose the correct name of the object using written or verbal choices, the patient has demonstrated comprehension of the therapist's question.

■ Present the patient with photos of famous or familiar people. Choose famous people whom the patient would likely know, or choose photos of the patient's family members. If the patient is unable to name the presented photos as a result of anomia, allow the patient a choice of three written or verbal answers provided by the therapist (the patient can shake his or her head "no"/"yes" in response to the choices).

■ If the patient can accurately choose the correct name of the person using written or verbal choices, the patient has demonstrated comprehension of the therapist's question.

☞ Anomia is indicated if the patient is unable to name three out of five presented familiar objects or is unable to name three out of five presented photos of famous people or family members.

3. *Agrommation* is the inability to arrange words sequentially so that they form intelligible sentences in conversation or writing. Agrommation usually occurs as a result of left hemisphere lesions. Patients mix up the correct sequence of words and fail to form clear, meaningful sentences. For example, a patient may write or say, "Fox over jumped fence chicken house," instead of, "The fox jumped over the chicken house fence." Or, "Mother cake oven bake,"

instead of "Mother baked the cake in the oven." Articles and prepositions such as "the" and "in" are left out. The therapist can observe the patient's speech during the screening to determine if agrommation is evident.

- Ask the patient to write (if possible) or repeat the sentences: "The fox jumped over the chicken house fence" and "Mother baked the cake in the oven."

☞ Agrommation is indicated if the patient mixes up the sequence of the words in these sentences (when written or verbally repeated).

4. *Agraphia* is the inability to write intelligible words and sentences. Agraphia is the written form of *alexia* (the inability to read) and occurs as a result of left hemisphere lesions. A patient with agraphia may attempt to write but instead only scribble. Agraphia can be observed by asking the patient to write his or her name, address, and phone number.

5. *Dysgraphia* is the inability to write because the patient cannot break down words into phonemes. Dysgraphia is the written form of *dyslexia* (the inability to read because the patient cannot break down words into phonemes) and occurs as a result of left hemisphere lesions. The writing of a patient with dysgraphia contains words with reversed letters, flipped letters, and a mixed-up sequence of letters. Some words may be left out of the sentence. Dysgraphia can be observed by asking a patient to write his or her name, address, and phone number. The attempts at writing by a patient with dysgraphia may appear as follows:

- Mray Smht (Mary Smith)
  5B6 Nil raod (56-B Mill Road)
  jnkimgont, PA 19964 (Jenkintown, PA 19046)

6. *Acalculia* is the inability to calculate mathematical problems. *Dyscalculia* is difficulty calculating math problems. Both are caused by lesions occurring in the left hemisphere. Although mathematical calculation is a cognitive function, it is considered to be a form of language expression using the language, or numerical symbols, of mathematics.

- Ask the patient to calculate simple addition and subtraction problems on paper or to make change in a simple word problem (e.g., The apple costs 75¢. You give the store employee $1.00. Show me what change in coins the employee will give you.).

☞ A patient with acalculia would be unable to perform mathematical functions.

In dyscalculia, the patient reverses numbers, flips them, mixes them up sequentially, or overlooks them. For example, if you ask a patient to write the addition problem of 16 plus 34, a patient with dyscalculia may write 19 + 43.

- Ask the patient to calculate simple addition and subtraction problems on paper or to make change in a simple word problem.

☞ A patient with dyscalculia would have difficulty performing mathematical functions as a result of reversing, flipping, overlooking, or sequentially mixing up the correct order or sequence of numbers.

MOTOR PERCEPTION DISORDERS

Motor perception disorders are referred to as *apraxias,* or motor planning problems. Apraxia can result from either right or left hemisphere lesions but usually results from right hemisphere lesions of the prefrontal area (or the anterior multimodal association area), the premotor area, and/or the primary motor cortex. Patients with apraxia have a distorted perception of the motor strategies required to negotiate their environment.

1. *Ideational apraxia* involves an inability to cognitively understand the motor demands of the task. For example, a patient may not understand that a shirt is an article of clothing to be worn on the torso and upper extremities.

2. *Ideomotor apraxia* involves the loss of the kinesthetic memory of motor patterns; in other words, the motor plan for a specific task may be lost. Or the motor plan may be intact, but the patient cannot access the appropriate motor plan and may implement an inappropriate motor plan for a specific task. For example, a patient may cognitively understand that a toothbrush is used for brushing one's teeth but may access an inappropriate motor plan for using a toothbrush and instead use the toothbrush to brush his or her hair. Sometimes a patient with an ideomotor apraxia cannot access a specific motor plan on command but can access it automatically when presented with a visual cue, such as offering a patient a comb.

Note: It is difficult to distinguish between an ideational and ideomotor apraxia. Screening methods for the apraxias are the same.

- ■ Apraxia can be detected by observing the patient carrying out motor sequences. Give the patient at least four of the following verbal commands:
  - Wave goodbye.
  - Blow a kiss.
  - Snap your fingers.
  - Touch your left knee.
  - Touch your right ear.
  - Cross your legs.
  - Raise your arm above your head.
  - Fold your arms across your chest.

☞ Apraxia is indicated if the patient has difficulty implementing the appropriate motor plan for three out of four of these motor tasks.

3. *Dressing apraxia* involves an inability to dress oneself due to body schema disorder or apraxia. With a *body schema disorder,* the patient's understanding of his or her body in space has become so distorted that dressing becomes extraordinarily difficult. For example, a patient with an attentional neglect syndrome (e.g., unilateral neglect, anosognosia) may dress only half of his or her body. With an *apraxia,* the patient (a) does not cognitively understand the demands of the task *(ideational apraxia),* (b) has lost the appropriate motor plan for a specific task *(ideomotor apraxia),* or (c) maintains the appropriate motor plan for a specific task but can no longer access it *(ideomotor apraxia).* For example, a patient with an ideomotor

apraxia who understands that pants are to be worn on the lower extremities may be unable to access the appropriate motor plan for donning pants and instead attempt to place his or her arms through the legs of his or her pants.

■ Observe the patient during morning activities of daily living. Apraxia is indicated if the patient has difficulty dressing; for example, the patient may inappropriately attempt to place the pants on his or her torso or may dress only half of his or her body. Apraxia also is indicated if the patient inappropriately uses the tools of grooming; for example, the patient uses his or her toothbrush to brush his or her hair.

■ Place objects in front of the patient and ask him or her to show you what one does with the object. State to the patient, "Show me what to do with this object."

■ Choose at least four of the following: a toothbrush, a comb, a folded letter and envelope, an article of clothing (slippers, a large shirt), a key, a wristwatch, a pen, a telephone.

☞ Apraxia is indicated if the patient is unable to demonstrate the appropriate motor plan for three out of the four presented objects.

4. *Two- and three-dimensional constructional apraxia* involves an inability to copy or build two- and three-dimensional designs. For example, a patient who is an architect would have difficulty drafting two-dimensional blueprints. A patient who is a plumber would have difficulty reassembling a kitchen faucet, once taken apart. Patients with constructional apraxia due to a right hemisphere lesion draw objects or put models together in a spatially disorganized way. Patients with constructional apraxia due to a left hemisphere lesion draw objects that lack detail. When building three-dimensional models, the models are spatially organized, but pieces are often left out. To determine whether constructional apraxia is present, choose one of the three screening procedures listed below.

■ Instruct the patient to construct a three-dimensional block design (using three-dimensional blocks) by copying a two-dimensional colored block design (shown on paper). Constructional apraxia may be indicated if the patient incorrectly constructs the three-dimensional block design.

■ Present the patient with a large bolt onto which is screwed a specific sequence of different-sized nuts and washers. Present the patient with a duplicate bolt and several sizes of nuts and washers. Instruct the patient to use the hardware to assemble an exact replica of the model bolt with its specific sequence of nuts and washers. Constructional apraxia may be indicated if the patient is unable to construct an exact duplicate of the model bolt.

■ Instruct the patient to construct a house (or car, person) using Legos™ (or similar toy construction pieces) based on a three-dimensional model provided by the therapist.

☞ Constructional apraxia may be indicated if the patient is unable to accurately construct a duplicate of the model provided by the therapist.

RED FLAGS
SIGNS AND SYMPTOMS OF PERCEPTUAL IMPAIRMENT

A red flag is an indicator or symptom of dysfunction. If a patient is observed displaying any of the following red flags, it is likely that the patient possesses perceptual deficits.

VISUAL PERCEPTION DISORDERS

- Unable to recognize objects and familiar people by sight alone, despite intact visual anatomy (visual agnosia, prosopagnosia).

- Distorts the physical properties of objects and people (metamorphopsia).

- Unable to interpret a visual stimulus as a whole (simultanagnosia).

- Unable to remember the names of colors (color anomia).

- Unable to remember the color of familiar objects (color agnosia).

VISUAL–SPATIAL PERCEPTION DISORDERS

- Confuses right and left (right–left discrimination disorder).

- Difficulty distinguishing objects in the foreground from objects in the background (figure–ground discrimination dysfunction).

- Unable to recognize familiar forms when they are viewed from an unfamiliar visual perspective (form–constancy dysfunction).

- Confuses concepts regarding positions in space—before/after, in/out, up/down, behind/in front of (position in space dysfunction).

- Frequently loses sense of direction and becomes lost (topographical disorientation).

- Misjudges the distances of objects in relation to oneself and the environment (depth perception dysfunction).

TACTILE PERCEPTION DISORDERS

- Unable to identify familiar objects by touch alone (astereognosis).

- Unable to discriminate between different types of materials by touch alone (ahylognosia).

- Unable to discriminate between different forms and shapes by touch alone (amorphagnosia).

- Unable to determine whether one has been touched by one or two points (two-point discrimination disorder).

- Unable to identify letters written on one's hand (agraphesthesia).

- Unable to feel sensation on one's involved side if simultaneously touched on the uninvolved side (extinction of simultaneous stimulation).

BODY SCHEMA PERCEPTION DISORDERS

- Unable to understand the spatial relationships between one's own fingers; unable to identify one's thumb, index, middle, ring, and little fingers (finger agnosia).

- Demonstrates inattention to one side of one's body or environment (unilateral neglect).

- Unable to recognize one's paralyzed limbs as belonging to one's own body (anosognosia).

- Unable to feel sensation on one's involved side if simultaneously touched on the uninvolved side (extinction of simultaneous stimulation).

LANGUAGE PERCEPTION DISORDERS: RECEPTIVE APHASIA

- Unable to comprehend spoken language, despite intact auditory anatomy (Wernicke's aphasia).

- Unable to read, despite intact visual anatomy (alexia).

- Unable to break words down into phonemes—reverses letters, flips letters, mixes up sequence of letters (dyslexia).

- Unable to comprehend familiar gestures and symbols (asymbolia).

- Unable to interpret the emotional tone of conversations (aprosodia).

LANGUAGE PERCEPTION DISORDERS: EXPRESSIVE APHASIA

- Difficulty expressing oneself through spoken language. Speech is characterized by nonfluent, telegraphic sentences. Substitutes inappropriate words for specific target words (Broca's aphasia).

- Unable to express the names of people or objects (anomia).

- Mixes up the sequence of words in spoken or written language (agrommation).

- Unable to write; demonstrates scribbling when asked to write (agraphia).

- Difficulty writing; writing characterized by reversal of letters, flipped letters, incorrect sequence of letters (dysgraphia).

- Difficulty using the language and symbols of mathematics (acalculia).

- Difficulty calculating mathematical problems; written math is characterized by number reversal, flipped numbers, and disarranged sequence of numbers (dyscalculia).

MOTOR PERCEPTION DISORDERS

- Unable to understand the motor demands of a specific task; for example, the patient does not understand what a shirt is for (ideational apraxia).

- Understands the motor demands for a specific task but is unable to access the motor plan for that specific motor task—the motor plan may be lost due to neurologic damage. The patient understands that a shirt is an article of clothing to be worn on the torso and upper extremities but does not know how to don the shirt (ideomotor apraxia).

- Accesses an inappropriate motor plan for a specific motor task. The patient understands that a shirt is an article of clothing to be worn on the torso and upper extremities but accesses the motor plan for donning pants. For example, the patient attempts to place his or her legs through the shirt sleeves (ideomotor apraxia).

- Unable to dress due to an ideational apraxia, an ideomotor apraxia, or a body schema perceptual disorder (dressing apraxia).

- Unable to construct two- and three-dimensional objects (two- and three-dimensional construction apraxia).

AVAILABLE IN-DEPTH ASSESSMENTS

If perceptual screening detects neurologic impairment, the following in-depth assessments are available:

Aphasia Diagnostic Profiles (ADP)
N. Helms-Estabrooks
Adults with acquired brain damage
Applied Symbolix
800 N. Wells St., Ste. 200
Chicago, IL 60610

Apraxia Battery for Adults (ABA)
B. L. Dabul
Adolescents and adults with motor planning deficits secondary to neurologic impairment
PRO-ED, Inc.
8700 Shoal Creek Blvd.
Austin, TX 78757-6897
(800) 897-3202, (512) 451-3246

Arnadóttir OT-ADL Neurobehavioral Evaluation (A-ONE)
G. Arnadóttir
Adults with cortical impairment
Arnadóttir, G. (1990). *The brain and behavior: Assessing cortical dysfunction through activities of daily living.* Philadelphia: Mosby.

Assessment of Motor and Process Skills (AMPS)
A. G. Fisher
Children and adults with developmental, neurologic, and musculoskeletal disorders
Anne G. Fisher, ScD, OTR/L, FAOTA
AMPS Project
Occupational Therapy Building

Colorado State University
Fort Collins, CO 80523

Boston Diagnostic Aphasic Examination, 3rd ed. (BDAE-3)
B. Goodglass, E. Kaplan, B. Barresi
Adults with aphasia secondary to neurologic insult
Psychological Corporation
555 Academic Ct.
San Antonio, TX 78204-2498
(800) 228-0752, (800) 211-8378

Behavioural Inattention Test (BIT)
B. Wilson, J. Cockburn, P. Halligan
Adults with visual neglect secondary to neurologic impairment
National Rehabilitation Services
117 North Elm St.
PO Box 1247
Gaylord, MI 49735
(517) 732-3866

Benton Visual Retention Test, 5th ed.
A. L. Benton
Children, adolescents, and adults with visual–perceptual processing impairment
Psychological Corporation
555 Academic Ct.
San Antonio, TX 78204-2498
(800) 228-0752, (800) 211-8378

Chessington OT Neurological Assessment Battery (COTNAB)
R. Tyerman, A. Tyerman, P. Howard, C. Hadfield, A. L. Benton

Adults with neurologic impairment
North Coast Medical
187 Stauffer Blvd.
San Jose, CA 95125-1042

Clock Test
H. Tuokko, T. Hadjistravropoulos, J. A. Miller,
A. Horton, B. L. Beattie
Elderly persons with dementia
Multi-Health Systems, Inc.
908 Niagara Falls Blvd.
North Tonawanda, NY 14120-2060
(416) 424-1700

Comprehensive Test of Visual Functioning (CTVF)
S. L. Larson, E. Buethe, G. J. Vitali
Children, adolescents, and adults with neurologic impairment
Slosson Educational Publications, Inc.
PO Box 280
East Aurora, NY 14052-0280
(800) 828-4800

Deuel's Test of Manual Apraxia
D. F. Edwards, R. K. Deuel, C. Baum, J. C. Morris
Adults with dementia
Dorothy F. Edwards, PhD
Washington University
4567 Scott Ave.
St. Louis, MO 63105

Elemental Driving Simulator (EDS) and Driving Assessment System (DAS)
R. Giatnutsos, A. Campbell, A. Beattie,
F. Mandriota
Adults with neurologic impairment
Life Sciences Associates
One Fenimore Rd.
Bayport, NY 11795-2115
(516) 472-2111

Examining for Aphasia, 3rd ed. (EFA–3)
J. Eisenson
Adolescents and adults with aphasia secondary to neurologic impairment
PRO-ED, Inc.

8700 Shoal Creek Blvd.
Austin, TX 78757-6897
(800) 897-3202, (512) 451-3246

Hooper Visual Organization Test (VOT)
H. E. Hooper
Adolescents and adults
Western Psychological Services
12031 Wilshire Blvd.
Los Angeles, CA 90025-1251
(800) 648-8857, (310) 478-2061

Katz Index of ADL
S. Katz, A. B. Ford, R. W. Moskowitz,
B. A. Jackson, M. W. Jaffe
Elderly persons with neurologic and/or musculoskeletal impairment
Katz, S., Ford, A. B., Moskowitz, R. W., Jackson, B. A., & Jaffe, M. W. (1963). The Index of ADL: A standardized measure of biological and psychological function. *Journal of the American Medical Association, 185*, 914–919.

Klein–Bell Activities of Daily Living Scale
R. M. Klein, B. J. Bell
Infants to adulthood—neurologic and/or musculoskeletal impairment
Health Sciences Center for Educational Resources
University of Washington
T-281 Health Science Bldg.
Box 357161
Seattle, WA 98195-7161

Kohlman Evaluation of Living Skills (KELS)
L. Kohlman-Thomson
Adolescents and adults with psychiatric, developmental, and/or neurologic impairment
American Occupational Therapy Association
4720 Montgomery Ln.
PO Box 31220
Bethesda, MD 20824-1220
(301) 652-2682

Minnesota Spatial Relations Test (MSRT)
R. V. Davis
Adolescents and adults with visual–spatial impairment

American Guidance Service, Inc.
Publishers' Bldg.
Circle Pines, MN 55014
(800) 328-2560

Minnesota Test for Differential Diagnosis of Aphasia
H. Schell
Adults with neurologic impairment
American Guidance Service, Inc.
4201 Woodland Rd.
Circle Pines, MN 55014-1796
(800) 328-2560

Motor-Free Visual Perception Test–Revised (MVPT–R) and Motor–Free Visual Perception Test–Vertical (MVPT–V)
MVPT–R: R. P. Colarusso, D. D. Hammill
MVPT–V: L. Mercier
Children and adults with visual–perceptual impairment
Academic Therapy Publications
20 Commercial Blvd.
Novato, CA 94947-6191
(800) 422-7249

OSOT Perceptual Evaluation, Revised (Ontario Society of Occupational Therapists)
P. Fisher, M. Boys, C. Holzberg
Adults with neurologic impairment
Nelson Canada
1120 Birchmont Rd.
Scarborough, ON M1K 5G4 Canada

Portable Tactual Performance Test (P-TPT)
Psychological Assessment Resources, Inc.
Children, adolescents, and adults with tactile agnosia
Psychological Assessment Resources, Inc.
PO Box 998
Odessa, FL 33556-9908
(800) 331-8378

Psycholinguistic Assessments of Language Processing in Aphasia (PALPA)
J. Kay, R. Lesser, M. Coltheart
Adults with aphasia
Psychology Press

325 Chestnut St., Ste. 800
Philadelphia, PA 19106
(215) 625-8900

Rivermead Perceptual Assessment Battery (RPAB)
G. Bhavnani, J. Cockburn, N. Lincoln, S. Whiting
Adults with neurologic impairment
Western Psychological Services
12031 Wilshire Blvd.
Los Angeles, CA 90025-1251
(800) 648-8857, (310) 478-2061

Sensory Integration Inventory–Revised, for Individuals with Developmental Disabilities
J. E. Reisman, B. Hanschu
Children and adults with developmental and/or neurologic impairment
PDP Products
12015 North July Ave.
Hugo, MN 55038
(612) 439-8865

Sklar Aphasia Scale, Revised (SAS)
M. Sklar
Adults with neurologic impairment
Western Psychological Services
12031 Wilshire Blvd.
Los Angeles, CA 90025-1251
(800) 648-8857, (310) 478-2061

Structured Observational Test of Function (SOTOF)
A. Laver, G. E. Powell
Elderly persons with neurologic impairment or dementia
NFER-NELSON Publishing Company
Darville House
2 Oxford Rd. East
Windsor, Berkshire SL4 1DF England
0753 858961

Test of Oral and Limb Apraxia (TOLA)
N. Helm-Estabrooks
Adults with neurologic impairment
Riverside Publishing Company
425 Spring Lake Dr.
Itasca, IL 60143-2079
(800) 767-8378, (800) 323-9540

Test of Visual–Motor Skills: Upper Level Adolescents and Adults (TVMS: UL)
M. F. Gardner
Adolescents and adults with neurologic impairment
Psychological and Educational Publications, Inc.
1477 Rollins Rd.
Burlingame, CA 94010
(800) 523-5775

Visual Search and Attention Test (VSAT)
M. R. Trenerry, B. Crosson, J. DeBoe, W. R. Leber
Adults with visual–perceptual deficits secondary to neurologic impairment
Psychological Assessment Resources, Inc.
PO Box 998
Odessa, FL 33556-9908
(800) 331-8378

For further information about the these assessments, refer to the following resources:

Asher, I. E. (1996). *Occupational therapy assessment tools: An annotated index* (2nd ed.). Bethesda, MD: American Occupational Therapy Association.

Impara, J. C., Plake, B. S., & Murphy L. L. (Eds.). (1998). *The thirteenth mental measurements yearbook.* Lincoln: University of Nebraska Press.

Murphy, L. L., Impara, J. C., & Plake, B. S. (Eds.). (1999). *Tests in print: An index to tests, test reviews, and the literature on specific tests* (Vols. 1 & 2). Lincoln: University of Nebraska Press.

Plake, B. S., Impara, J. C., & Murphy, L. L. (Eds.). (1999). *The supplement to* The Thirteenth Mental Measurements Yearbook. Lincoln: University of Nebraska Press.

PERCEPTUAL SCREENING

Patient Name _____   Date _____

Perceptual Skills	No Impairment Detected	Possible Impairment Detected	Comments
VISUAL PERCEPTION			
Visual agnosia			
Prosopagnosia			
Simultanagnosia			
Metamorphopsia			
Color agnosia			
Color anomia			
VISUAL–SPATIAL PERCEPTION			
Right–left discrimination			
Figure–ground discrimination			
Form–constancy discrimination			
Topographical disorientation			
Depth perception			
TACTILE PERCEPTION			
Astereognosis			
Ahylognosia			
Amorphagnosia			
Two-point discrimination			
Agraphesthesia			
Extinction of simultaneous stimulation			
BODY SCHEMA PERCEPTION			
Finger agnosia			
Unilateral neglect			
Anosognosia			
Extinction of simultaneous stimulation			
LANGUAGE PERCEPTION			
Receptive aphasia			
Wernicke's aphasia			
Alexia/dyslexia			
Asymbolia			
Aprosodia			
Expressive aphasia			
Broca's aphasia			
Anomia			
Agrommation			
Agraphia			
Acalculia/dyscalculia			
MOTOR PERCEPTION			
Ideational apraxia			
Ideomotor apraxia			
Dressing apraxia			
Two- and three-dimensional apraxia			

SECTION 4

Sensory SCREENING

FUNCTIONAL IMPLICATIONS OF SENSORY DISORDERS

The *sensory system* is vital for learning and exploration of our environment, achieving functional motor performance, and providing safety and protection of the organism. A strong relationship exists between *sensory input* and *motor output*. For example, a patient who has decreased awareness of his or her limbs in space will have great difficulty or an inability to reach for an item in a paper bag or execute appropriate motor responses to an environmental demand.

Sensory evaluation allows therapists to

- Assess the extent of sensory loss;

- Identify lesion location;

- Evaluate and document sensory recovery;

- Determine functional impairment and limitations; and

- Guide treatment intervention.

Two sensory systems provide information about our internal and external environments. They are the *somatosensory* and *special sensory systems*. The somatosensory system provides our most basic sensory information and can be subdivided into two components:

- *Primary somatosensory system* includes light touch, pain, temperature, pressure, vibration, proprioception, and kinesthesia. For example, the primary somatosensory system allows an individual to distinguish when tap water has become too hot or when a scalp hair, too light to see, is on our arm. The primary somatosensory system is mediated by the skin receptors, spinal nerves, spinal cord tracts, and the cortical somatosensory area 1 (SS1).

- *Cortical* or *secondary somatosensory system* includes two-point discrimination, stereognosis, graphesthesia, and extinction of simultaneous stimulation. This system provides complete integrated information that is perceptual rather than receptive. For example, this system allows one to guide one's fingers into a glove, reach for a pen within a knapsack, or feel the vibration of a booting-up computer. The cortical secondary somatosensory system is mediated by the secondary somatosensory cortex (SS2) and the posterior multimodal association cortex.

The special sensory system comprises vision, olfaction, gustation, audition, and equilibrium. For example, this system enables one to be alerted to food burning on a stove or to changes in a support surface, such as a slope in the sidewalk. We have various types of sensory receptors:

- *Exteroceptors* respond to external stimuli such as those of our special senses. They allow individuals to distinguish between spicy and sweet foods.

- *Interoceptors* respond to stimuli provided by our internal organs. For example, a patient with left arm and jaw pain is alerted to the first signs of a heart attack.

- *Proprioceptors* are located in the muscles, tendons, ligaments, joints, and inner ear. They detect changes in body position and movement. These receptors allow one to reach for a glass on a top shelf when vision is occluded.

Sensory receptors vary in anatomical location:

- *Cutaneous receptors* lie within the superficial and deep layers of the skin. Pathology may inhibit a stroke patient with hemisensory loss to feel pain if he or she accidentally contacts the hot stove during meal preparation.

- *Proprioceptors* are receptors located within the muscles, tendons, and joints. A patient with hemisensory loss may accidentally entangle his or her affected arm within the spokes of a wheelchair.

- *Visceral receptors* respond to pain from the internal organs. A patient experiencing left arm and jaw pain is alerted to the first signs of a heart attack.

- The *primary somatosensory cortex* is located in the postcentral gyrus. It is topographically organized to represent all sensory areas within the body.

- Sensory receptors within the *peripheral nervous system* transmit sensory information via the dorsal root to the dorsal rootlets. Within the dorsal horn, the receptors synapse on sensory spinal cord tracts. Sensory information is further transmitted upward to the brain stem and to the thalamus. Within the thalamus, information is organized and relayed to the cortex, where it is detected and interpreted.

- The *cerebellum* plays an additional role in information analysis of proprioceptive data received from the muscles and joints. The sensory information received by the cerebellum is used to achieve coordinated body movements and balance.

- Specific skin segments are innervated by spinal nerves. These regions are called *dermatomes*. Clinically, these specific sensory divisions assist therapists to determine sensation of a specific region and/or lesion location.

Disease or injury/trauma to the sensory receptors, spinal nerves, spinal cord tracts, brain stem, thalamus, and cortical areas contribute to sensory pathology. Patients may experience *hypersensitivity,* making most tactile stimuli noxious or painful. *Hyposensitivity* leads to a partial or incomplete ability to detect sensory stimuli or accurately interpret sensations.

Somatosensory and special senses are transmitted via seven *sensory spinal cord tracts*. Patients with *dorsal column lesions* above the brain stem have contralateral sensory deficits involving (a) discriminatory touch, (b) pressure, (c) pain, (d) proprioception, and (e) kinesthesia. Activities of daily living (ADL) tasks that require movement with occluded vision, such as fastening a necklace, placing one's hair in a ponytail holder, or reaching into a pocket for a quarter, become challenging. Additionally, patients may become at risk for skin breakdown on the involved extremity due to an ill-fitting orthotic device.

Patients with *lateral thalamic tract lesions* above the brain stem present with deficits in contralateral hemisensory loss for (a) pain and (b) temperature. Kitchen management tasks such as washing dishes or stovetop activities may be hazardous due to a decreased ability to distinguish between hot and cold water temperatures. Prolonged exposure of the involved extremity to cold temperatures may increase a patient's risk for frostbite.

Patients with *anterior spinothalamic tract lesions* above the brain stem present with deficits in contralateral hemisensory loss for (a) light touch and (b) crude touch. Feeding, dressing, and grooming/hygiene tasks may become frustrating as the patient regularly drops items from the involved extremity.

Patients with *cuneocerebellar tract* and *rostral spinocerebellar tract lesions* will demonstrate deficits in ipsilateral upper-extremity and trunk proprioception. Patients may exhibit deficits in trunk control, difficulty maintaining midline position, difficulty initiating weight shifts during transfers, and clumsiness during reaching activities.

Patients with *cranial nerve lesions* of the *olfactory nerve* (cranial nerve 1), *optic nerve* (cranial nerve 2), *trigeminal nerve* (cranial nerve 5), *facial nerve* (cranial nerve 7), *vestibulocochlear nerve* (cranial nerve 8), *glossopharyngeal nerve* (cranial nerve 9), and *vagus nerve* (cranial nerve 10) may demonstrate functional impairment in the above-mentioned special sensory system:

- *Involvement of cranial nerve 1 impairs smell and taste.* For example, one might be unable to detect spoiled food, thus increasing one's risk for gastrointestinal illnesses, or one may be unable to detect the odor of burning electrical wires, thus increasing potential fire hazards.

- *Involvement of cranial nerve 2 impairs visual acuity, peripheral vision, and color detection.* For example, a patient may experience difficulty reading the small print on a medication bottle. Patients may have difficulty selecting clothing of similar hues (brown, black, blue). Eating a light-colored entrée on a light-colored plate may become a difficult task, thus limiting a person's desire for social dining. Patients with a homonymous hemianopsia may require extra caution when negotiating streets or shopping in busy malls because of a narrow scanning pattern.

- *Involvement of cranial nerve 5 impairs sensation for the face, head, oral cavity, and cornea.* Patients may become at risk for corneal abrasions leading to refractive errors due to decreased or absent sensation of the cornea. Social dining may be affected, as a patient with sensory loss around the oral area is unable to detect food on his or her lips or face. Patients with hemisensory loss of the oral cavity may pocket food in their mouth, aspirate, and develop aspiration pneumonia.

- *Involvement of cranial nerve 7 alters the ability to detect sweet substances* on the anterior portion of the tongue. For example, a patient may have difficulty determining how much sugar to add during a baking activity.

- *Involvement of cranial nerve 8 impairs hearing and equilibrium.* Increased precaution may be needed when crossing the street, as the patient is unable to localize the direction of a siren. Patients may experience vertigo or dizziness during a simple task such as reaching down to retrieve a shoe or rolling in bed.

- *Involvement of cranial nerve 9 impairs the ability to detect sour food,* such as spoiled milk.

- *Involvement of cranial nerve 10 impairs visceral control,* which may impede regulation of heart rate, digestion, or smooth muscle control.

Patients with *vascular lesions* will present with sensory loss in the following vascular distributions:

- *Anterior cerebral artery lesions* will result in contralateral lower-extremity sensory loss. Functionally, patients may experience difficulty with ambulation, negotiation of stairs, and bathroom transfers.

- *Middle cerebral artery lesions* will result in contralateral sensory loss of the upper-extremity, lower-extremity, and facial regions. Functionally, a patient may have difficulty identifying objects in his or her pockets. Caution should be taken when drinking and eating hot foods to prevent superficial burns to the lips and mouth.

- *Lenticulostrate and thalamoperforate arterial lesions* result in contralateral sensory loss to one entire side of the body. Functionally, these patients will be challenged with postural symmetry and the coordination of movement necessary for basic ADL tasks.

Patients with *subcortical lesions* including the internal capsule, thalamus, and parietal lobe will demonstrate contralateral sensory loss involving the face, upper extremities, and lower extremities. Lesions involving the thalamus present with sensory loss of the face, upper extremities, lower extremities, and the trunk.

Patients with *parietal lobe lesions* exhibit difficulty with sensory motor integration. Functionally, patients may exhibit difficulty selecting a key and opening the door or manipulating a remote control when watching television.

Cortical lesions affect the patient's ability to interpret meaningful sensory information. For example, a patient may have difficulty recognizing different types of coins in his or her hand (astereognosis) or recognizing numbers written in the palm of his or her hand (agraphesthesia). These deficits are associated with cortical sensory loss.

SENSORY SCREENING PROCEDURES

Before evaluating the patient, perform a thorough chart review, noting lesion or injury location, if available. All patients who have sustained a neurological disease or injury should be screened for sensory dysfunction. Knowledge of the patient's cognitive, hearing, and visual acuity status is necessary for screening administration. Refer to the cognitive screening section of this text for orientation arousal, recognition, attention, memory, comprehension, and the ability to follow directions.

It is necessary to establish an alternate response system if the patient is not verbal. It is essential to explain the goals and purpose of the sensory screening to the patient. Always explain the screening procedures to the patient and ask for feedback to ensure that the instructions are understood. The therapist should emphasize cooperation from the patient, as it is necessary to ensure the most accurate results.

The testing environment should be a quiet room with minimal distractions. The patient should be comfortable and relaxed. Sitting upright in a chair or bed is the most optimal position, as it maximizes the patient's arousal level. A quick visual search for scar tissue and callus formations should be conducted because these conditions could contribute to decreased sensory status. The patient should wear loose-fitting clothing, making all areas accessible.

The therapist should perform a trial of each specific screening procedure with the patient's eyes open before the actual test in order to minimize anxiety. During the actual screening, the patient's eyes must be occluded either by having the patient voluntarily close them or having the therapist provide a shield or a blindfold. The therapist should allow the patient to open his or her eyes between each specific sensory screening.

Although most sensory screening results are assessed and documented through dermatome areas, this functional sensory screening is guided by functionality. Assessment of specific areas includes facial, oral, upper extremities, hands, trunk, lower extremities, and feet. The screening stimuli should be presented proximally to distally and on dorsal regions first, followed by ventral regions. The screening should begin on the nonaffected side, although with various lesions, sensory loss may be present on the ipsilateral side. Stimuli should be presented in a random, unpredictable manner to prevent perceived learning by the patient. For example, when assessing temperature, the therapist should apply one hot stimulus, one cold stimulus, and then no stimulus. When assessing discriminative stimuli, the therapist should alternate among sharp, dull, and no stimulus. A time limit of 10 to 15 seconds should be allowed between trials. Ask the patient to respond to the presented stimulus with the established communication system. Record the screening results on the form provided at the end of this section.

The Warren Near Text Card, the Lea Symbols Near Vision Test Card, and the Lighthouse Near Acuity Text Card are designed specifically for the low-vision population and may be purchased through Mattingly International (contact information can be found at the end of this section).

EVALUATION OF THE SPECIAL SENSES

1. *Olfaction (smell).* The function of smell is controlled by the olfactory nerve.

- Screen one olfactory nerve at a time.

- Occlude the patient's vision.

- Block the patient's nostril on the opposite side being tested.

- Present one odor at a time. Present at least four familiar scents, such as peppermint, cinnamon, coffee, and vanilla. Ammonia and acetic acid should be avoided because their pungency can stimulate nerve endings in the mucous membranes, giving a false-positive result. The patient may think that he or she is smelling the odor when in fact he or she is tasting the odor's dissolving vapors.

- If the patient is unable to name a specific odor, provide him or her with multiple verbal choices of the screened odors.

☞ An olfactory nerve impairment is indicated if the patient has difficulty smelling and identifying three out of the four presented odors.

2. *Vision.* The function of vision is mediated by the optic nerve and is responsible for visual acuity. The interpretation of visual information is mediated by the occipital lobe. Functional acuity is the ability to see detail for both near and far distances. Near distance is that associated with reading distance and performing near tasks such as reading, needlepoint, buttoning, and writing. Far distance is that associated with tasks such as driving, reading road signs, discriminating faces and objects, and watching television.

Near distance:

- In a well-illuminated room, place the patient in a seated position.

- Provide the patient with a near visual acuity chart (Warren Near Text Card or Lighthouse Near Acuity Text Card) held 16 inches away from the eyes (see Figure 2.8).

- Have the patient read from top to bottom.

- Continue until the patient misses more than 50% of the line or reading speed is considerably decreased.

Functional task:

- Ask the patient to read a line of standard-sized print in a newspaper or magazine.

- Continue for one page.

☛ A visual impairment is indicated if the patient complains of blurry or fuzzy print, is unable to bring the printed material into focus, complains that the print size is too small, or changes the focal length of the reading material. The patient should be able to read at an acuity level of 20/40 or better. Impairment may suggest pathology to the anterior chamber of the eye, optic nerve involvement, macular scotoma, vitreous hemorrhage, occipital lobe involvement, decreased contrast sensitivity, or decreased pupillary control.

Note: The patient should wear corrective reading glasses during the screening task, if applicable. The patient should use binocular vision and not adjust the focal length of the page during the task. If the patient is unable to read letters secondary to aphasia, the Lea Symbols Near Vision Test Card can be used. It is important to have the patient read for 3 to 5 minutes or the length of one page to observe for signs of fatigue, decreased concentration, and decreased oculomotor control. The therapist should also observe for omission of letters and words, losing place within text, slow reading speed, and a change in head positioning. If the patient appears confused by the lines or letters, cover all other lines not being read.

Far distance:

- In a well-illuminated room, place the patient in a seated position.

- Provide the patient with a distance acuity chart (Colenbrander or Snellen Acuity Chart) (see Figure 2.9). Chart distance should be 10 or 20 feet from the patient, depending on which chart is being used.

- Occlude the patient's left eye with an eye patch. Ask the patient to verbally identify the letters in the specified line.

- Have the patient read top to bottom and left to right.

- Continue until the patient misses more than 50% of the letters in one line.

- Record acuity as the last line in which the patient was able to read more than 50% of the letters. Note the number of letters missed; for example, 20/40-2 is defined as two out of five numbers missed when reading this particular ototype.

- Repeat with the right eye occluded.

- Repeat with both eyes (binocular vision).

Functional task:

- Have the patient read signs and room numbers either during functional ambulation or when navigating at wheelchair level.

☛ A visual impairment is indicated if the patient reads at an acuity level less than 20/40, complains of blurry or fuzzy words, complains of difficulty bringing print/words into focus, or attempts to move closer to the material being read.

Note: The patient should wear corrective distance lenses if applicable. If the patient is unable to identify the letters or the numbers on the acuity chart, substitute with the Lea Symbols. The patient should be instructed to maintain a stable head position and focal length/distance throughout the entire test. The therapist should also observe for omission of words or letters and any signs of changing head position to enhance acuity. If the patient appears confused by the lines or letters, cover all other lines not being read.

3. *Gustation (taste).* The function of taste is mediated by the facial nerve, which is responsible for the taste receptors on the anterior tongue, and the glossopharyngeal nerve, which is responsible for the taste receptors on the posterior aspect of the tongue. Each part of the tongue can detect all tastes; however, the anterior portion primarily detects sweet tastes, and the posterior primarily detects bitter tastes.

Screening the sensory portion of the facial nerve—screen for the sense of taste on the anterior tongue:

- Place the patient in a seated position with vision occluded.

- Present sweet, salty, and sour solutions to the lateral portions of the anterior tongue using a cotton applicator.

- Apply the taste substance first to the uninvolved side of the tongue, then to the involved side.

- Present each taste substance one at a time. Use a separate cotton applicator for each taste substance.

- Ask the patient to indicate whether he or she can taste the substance and identify whether it is sweet, sour, or salty.

☞ Impairment is indicated if the patient is unable to detect taste on one or both sides of the anterior tongue.

Screening the sensory portion of the glossopharyngeal nerve—screen for the sense of taste on the posterior tongue:

- Place the patient in a seated position with vision occluded.

- Present a bitter solution to the posterior aspect of the tongue (e.g., rind of a lemon).

- Use a cotton-tipped wooden applicator.

- Apply the taste substance first to the uninvolved side of the tongue, then to the involved side.

- Ask the patient to indicate whether he or she can taste the substance and identify what the taste substance is. Ask the patient to identify whether the taste substance is sweet, sour, salty, or bitter.

☞ Impairment is indicated if the patient is unable to detect taste on one or both halves of the posterior region of the tongue.

4. *Audition, balance, and equilibrium.* The vestibulocochlear nerve mediates the function of audition, balance, and equilibrium. The vestibulocochlear nerve has two branches: the auditory branch and the vestibular branch.

Screening auditory branch function:

- Strike a tuning fork and place it against the middle of the patient's forehead (Weber Test). Sound is heard through bone conduction and should be heard equally in both ears (see Figure 4.1). If sound is heard more loudly in one ear, this may indicate a conductive hearing loss.

- Strike a tuning fork and place it 1 inch from the auditory canal (Rinne Test) (see Figure 4.2).

- Place the vibrating tuning fork against the mastoid bone.

- Alternate placement until no sound is heard by the patient. Sound conducted by air is heard longer than sound conducted by bone.

Figure 4.1 Tuning Fork on Forehead: Screening Auditory Branch of Vestibulocochlear Nerve

☞ If the patient hears the sound longer when the tuning fork is placed on the mastoid bone, sensorineural hearing loss is indicated.

Screening vestibular branch function—screen for the presence of nystagmus:

Figure 4.2 Tuning Fork on Mastoid Bone (Auditory Canal): Screening Avidory Branch of Vestibulocochlear Nerve

- Place the patient in a seated position.

- Ask the patient to maintain his or her head in a fixed position while visually tracking a moving stimulus (a pen cap) held at a distance of 15 inches from the patient's face.

- Move the pen cap in an *H* and an *X* pattern.

- Check for nystagmus both within and at the end ranges of the visual fields.

☞ Impairment is indicated if nystagmus is detected in the temporal or nasal visual fields.

Screening for balance and for the presence of protective responses (Rhomberg Test):

- Ask the patient to assume a standing position with eyes open.

- Check for increased sway and loss of balance.

- Gently displace the patient's balance (see Figure 4.3).

- Check for the appropriate presence of protective responses.

- Repeat the procedure with the patient's eyes closed.

☞ Impairment is indicated if the protective responses are absent or sluggish.

Screening for the presence of extensor tone in the lower extremities:

- Attempt to take the patient's lower extremity through its full range of motion.

- Test one lower extremity at a time. Note whether extensor tone is present at the hip, knee, and ankle joints.

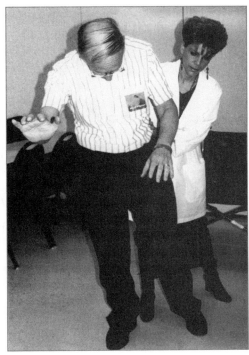

Figure 4.3 Presence of Protective Responses: Gently Displace the Patient's Balance

- Perform quick stretches to the hip, knee, and ankle to elicit the presence of extensor tone (see "Screening Muscle Tone" in Section 5).

☞ Increased tone in the lower extremities may indicate impairment in the vestibular branch of the vestibulocochlear nerve.

EVALUATION OF THE PRIMARY SOMATOSENSORY SYSTEM

1. *Light touch.* The function of light touch is to alert the individual of contact to one's skin.

- Occlude the patient's vision.

- Stroke lightly a small region of the patient's skin using a cotton swab or fingertip (see Figure 4.4).

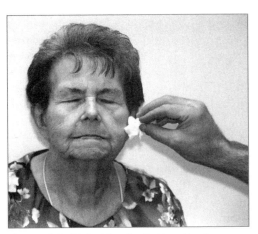

Figure 4.4 Light Touch: Cotton Swab

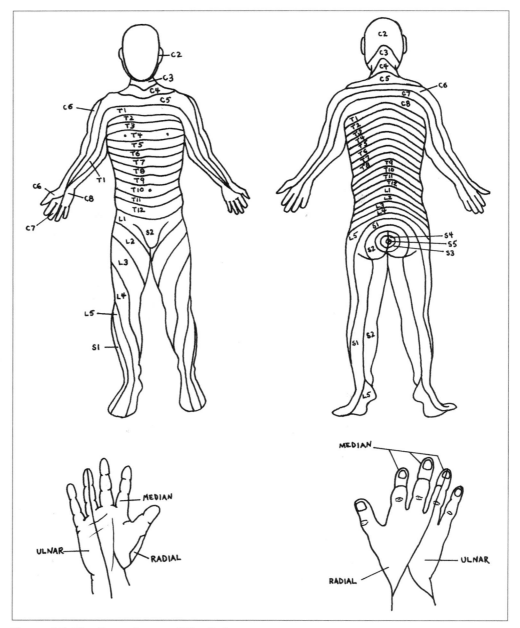

Figure 4.5 The Dermatomes of the Body

- Apply the stimulus to the patient's face, upper arm, forearm, digits, and toes, in accordance with the Dermatome Chart (see Figure 4.5).

- Apply the stimulus three times to a specific area.

- Ask the patient to indicate yes or no when the stimulus is felt. If yes or no verbalizations are not possible, the patient should respond with an alternative established communication method.

■ Document whether the patient presents with sensory loss along specific dermatome regions.

Light touch functional assessment:

■ Can the patient feel the shirt material on his or her arm?

■ Can the patient feel the sock sliding over his or her foot?

☞ Impairment in light touch is indicated if the patient is unable to identify when the stimulus was applied. A score of two out of three incorrect answers is indicative of impaired light touch.

2. *Pain.* The awareness of pain on the surface of the skin is a primitive protective measure alerting the organism of potential danger.

■ Occlude the patient's vision.

■ Randomly apply either the dull or sharp end of a safety pin or paper clip to the screening area (see Figure 4.6).

■ Apply the stimulus to the patient's face, upper arm, forearm, digits, bottom of the feet, and toes, in accordance with the Dermatome Chart (see Figure 4.5).

■ Apply the stimulus three times to a specific area.

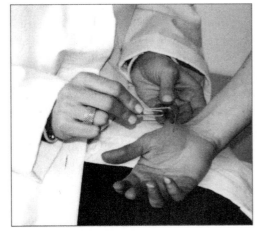

Figure 4.6 Sharp/Dull

■ Ask the patient to indicate sharp or dull when the stimulus is felt. If sharp or dull verbalizations are not possible, the patient should respond with the established alternative communication method.

■ Document whether the patient presents with sensory loss along specific dermatome regions.

Note: Do not use a safety pin if the patient presents with skin breakdown; open wounds; or dry, cracked skin.

Pain functional assessment:

■ Can the patient feel or identify sharp items or utensils when reaching into a kitchen drawer?

■ Can the patient feel or identify a shoe placed on the wrong foot?

☞ Impairment in pain sensation is indicated if the patient is unable to identify pain sensation when applied. A score of two out of three incorrect answers is indicative of impaired pain sensation.

3. *Temperature.* The ability to distinguish various changes in temperature serves as a protective measure.

- Occlude the patient's vision.

- Randomly apply test tubes containing hot and cold water.

- Apply the stimulus to the patient's face, upper arm, forearm, trunk, digits, lower extremities, bottoms of the feet, and toes, in accordance with the Dermatome Chart (see Figure 4.5).

- Apply the stimulus three times to a specific area.

- Ask the patient to indicate hot or cold when the stimulus is felt. If hot or cold verbalizations are not applicable, the patient should respond with the established alternative communication method.

- Document whether the patient presents with sensory loss along specific dermatome regions.

Temperature functional assessment:

- Is the patient aware of the changing water temperature when washing his or her hands?

- Can the patient distinguish the surface temperature on an electric stovetop?

- Can the patient detect the water temperature when stepping into the bathtub?

☞ A sensory impairment is indicated if the patient is unable to distinguish hot from cold or if the patient is unable to recognize and accurately interpret temperature stimuli.

4. *Proprioception* is the ability to identify one's trunk and limb position in space when vision is occluded. *Kinesthesia* is the ability to identify one's limbs as they move through space when vision is occluded. Proprioception helps to maintain one's balance during static and transitional movements, motor control, and all aspects of ADL. *Note:* The therapist must hold the joint segment being assessed at the lateral aspects. This will prevent the patient from receiving cutaneous input.

- Occlude the patient's vision.

- Move the patient's involved upper extremity three times between shoulder flexion and extension. The last position should place the extremity into shoulder flexion.

- Ask the patient, "Is your arm up or down? Can you replicate the position with your other arm?"

- Move the patient's involved upper extremity three times between horizontal abduction and adduction. The last position should place the extremity into horizontal abduction. Ask the patient, "Is your arm out to the side or next to your body? Can you replicate this position with your other arm?"

- Move the patient's involved elbow three times into flexion and extension. The last position should place the elbow into extension. Ask the patient, "Is your elbow bent or straight? Can you replicate this position with your other arm?"

- Move the patient's involved forearm three times between supination and pronation. The last position should place the forearm into supination. Ask the patient, "Is your hand facing up toward the ceiling or down toward the floor? Can you replicate this position with your other arm?"

- Move the patient's involved wrist three times between flexion and extension. The last position should place the wrist into flexion. Ask the patient, "Is your wrist bent or straight? Can you replicate this position with your other wrist?"

- Move the patient's involved thumb three times between thumb flexion and extension. The last position should place the thumb into extension. Ask the patient, "Is your thumb bent or straight? Can you replicate this position on your other hand?"

- Move the patient's involved fifth digit three times between flexion and extension. The last position should place the digit into flexion. Ask the patient, "Is your pinky bent or straight? Can you replicate this position on your other hand?"

- Move the patient's involved knee three times between flexion and extension. The last position should place the knee into flexion. Ask the patient, "Is your knee bent or straight? Can you replicate this position on your other leg?"

- Move the patient's involved big toe three times between flexion and extension. The last position should place the big toe into flexion. Ask the patient, "Is your big toe bent or straight? Can you replicate this position on your other foot?"

☞ A sensory impairment is indicated if the patient is unable to state the position of the assessed joint or if the patient is unable to reproduce the joint position within 15° of the tested position.

Note: The patient should demonstrate the reproduced position on the noninvolved extremity within 15° of the involved extremity being assessed.

5. *Tactile localization* provides individuals with the ability to localize touch sensation on the skin area. Functionally, the ability to localize touch allows the individual to feel clothing on

his or her body, identify and remove the annoying fly on his or her arm, and localize a pressure spot from a shoe.

■ Occlude the patient's vision.

■ Use the Dermatome Chart as a guide for the following instructions (see Figure 4.5).

■ Apply a light touch using either a fingertip or an esthesiometer to the patient's uninvolved upper arm for 1 to 2 seconds. Ask the patient to touch the exact spot stimulated with her other hand (see Figure 4.7).

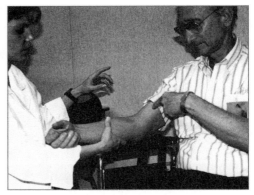

Figure 4.7 Tactile Localization

■ Apply a light touch using the same technique as above to the patient's uninvolved forearm. Ask the patient to touch the exact spot stimulated with his or her other hand.

■ Apply a light touch using the same technique as above to the patient's uninvolved palm. Ask the patient to touch the exact spot stimulated with his or her other hand.

■ Repeat the same procedure on the involved extremity.

■ Apply a light touch using the same technique as above to the patient's uninvolved thigh. Ask the patient to touch the exact spot stimulated with his or her other hand.

■ Apply a light touch using the same technique as above to the patient's uninvolved dorsal aspect of the foot. Ask the patient to touch the exact spot stimulated with his or her other hand.

■ Repeat the same procedure on the involved extremity.

■ Document whether the patient presents with sensory loss along specific dermatome regions.

☞ A sensory impairment is indicated if the patient is unable to localize the tactile stimulation within 1 inch of the actual stimulated area.

Note: If the patient is unable to touch the stimulated body part or location with the affected extremity, the patient can verbally tell the therapist the exact location of stimulation.

6. *Vibration* is the ability to perceive a vibratory stimulus. Functionally, the ability to perceive vibration alerts the individual to changes in the environment and signs of danger.

- Occlude the patient's vision.

- Strike a 30-cps tuning fork and apply the vibrating stimulus to the patient's uninvolved elbow (olecranon). It is important that the stem of the tuning fork is applied to the area being screened.

- Ask the patient to indicate when he or she feels the vibration.

- Strike a tuning fork and apply the vibrating stimulus to the patient's uninvolved knee (patella).

- Ask the patient to indicate when he or she feels the vibration.

- Repeat the above procedure with the involved extremities.

Vibration functional assessment:

- Can the patient feel the vibration of a stereo speaker with his or her hand?

- Can the patient feel the vibration of a computer booting up with his or her hand?

☞ A sensory impairment is indicated if the patient is unable to identify any of the vibratory stimuli indicated above.

EVALUATION OF THE CORTICAL OR SECONDARY SOMATOSENSORY SYSTEM

1. *Two-point discrimination* allows one to determine whether he or she has been touched by one or two points. This sensory function allows an individual to perform precision grip movements during ADL and instrumental activities of daily living tasks because it prevents one from dropping an object.

- Occlude the patient's vision.

- Apply an esthesiometer to the patient's uninvolved thumb (see Figure 4.8). Ask the patient whether one or two points have touched him or her. Begin with two distinct points and gradually move the distance closer. The starting distance of the esthesiometer should begin at 4 mm and end at 2.6 mm.

- Apply an esthesiometer to the patient's uninvolved index finger. Ask the patient whether one or two points have touched him or her.

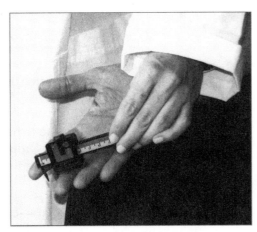

Figure 4.8 Two-Point Discrimination Using an Esthesiometer

Begin with two distinct points and gradually move the distance closer. The starting distance of the esthesiometer should begin at 4 mm and end at 2.6 mm.

■ Apply the esthesiometer to the patient's uninvolved fifth digit. Ask the patient whether one or two points have touched him or her. Begin with two distinct points and gradually move the distance closer. The starting distance of the esthesiometer should begin at 4 mm and end at 2.5 mm.

■ Repeat the same screening procedure on the involved hand.

☞ A sensory impairment is indicated if the patient is unable to discriminate between the above-stated millimeter distances for each tested area.

Note: If the patient is unable to verbally communicate the number of points stimulated, the patient can hold up the correct number or fingers or use a head nod.

2. *Stereognosis* is the ability to recognize shapes, objects, and textures by touch, with vision occluded. Functionally, stereognosis is one of the most important discriminative senses humans have because it allows individuals to identify familiar objects without the use of vision. For example, one can discern a quarter from a nickel when reaching into a pocket for change.

■ Occlude the patient's vision.

■ Select six familiar objects, such as a coin, fork, screw, key, pen, and toothbrush.

■ Place one object at a time within the patient's involved hand. Begin with the uninvolved extremity. It is important to ensure that the patient manipulates the object using the thumb and fingers.

■ Allow the patient to manipulate the object for 5 to 10 seconds.

■ Ask the patient to name the manipulated object.

■ Repeat with the involved extremity.

☞ Astereognosis is indicated if the patient is unable to identify three or more of the objects.

Note: If the patient is unable to name the object secondary to aphasia, it is acceptable to allow the patient to describe the properties of the object or to point to a selection of displayed choices identical to the testing items.

Astereognosis can be further broken down into two distinct skills: ahylognosia and amorphagnosia. *Ahylognosia* is difficulty or inability to identify textures and materials by touch alone. *Amorphagnosia* is difficulty or inability to identify shapes by touch alone.

Ahylognosia:

- Occlude the patient's vision.

- Select five familiar textures and materials (e.g., velvet, sandpaper, high-pile carpeting, smooth glass or plastic, burlap).

- Place one texture at a time within the patient's hand. Begin with the uninvolved extremity. It is important to ensure that the patient manipulates the texture using the thumb and fingers.

- Allow the patient to manipulate the texture and material for 5 to 10 seconds.

- Ask the patient to name the manipulated texture.

- Repeat with the involved extremity.

☞ Ahylognosia is indicated if the patient is unable to identify three or more of the textures.

Amorphagnosia:

- Occlude the patient's vision.

- Select five familiar shapes (e.g., round ball, square box or block, cylinder or soda can, triangle or pyramid, intact eggshell).

- Place one shape at a time within the patient's hand. Begin with the uninvolved extremity. It is important to ensure that the patient manipulates the shape using the thumb and fingers.

- Allow the patient to manipulate the shape for 5 to 10 seconds.

- Ask the patient to name the manipulated shape.

- Repeat with the involved extremity.

☞ Amorphagnosia is indicated if the patient is unable to identify three or more of the shapes.

3. *Graphesthesia* is the ability to interpret letters written on the palmar surface of one's hand.

- Instruct the patient that you are going to write letters on his or her hand with your fingertip.

- Write five letters on the patient's hand, one at a time (e.g., *o, w, t, x, s*). Ask the patient to identify each letter.

- Test the uninvolved hand first.

- Repeat with the involved hand.

☞ A sensory impairment is indicated if the patient is unable to identify three out of five letters.

4. *Extinction of simultaneous stimulation.* Simultaneous stimulation allows one to perceive touch on both sides of the body. This sensory function allows individuals to integrate different types of movement and to use bimanual manipulation during all aspects of ADL tasks.

- Touch the patient on the dorsal surface of the involved hand.

- Ask the patient to indicate on what body region he or she has been touched.

- Touch the patient on the dorsal surface of the uninvolved hand.

- Ask the patient to indicate on what body region he or she has been touched.

- Simultaneously touch the patient on the dorsal surface of both the uninvolved and involved dorsal hand.

- Ask the patient to indicate on what body region he or she has been touched.

- Repeat the above procedures with all extremities.

☞ Extinction of simultaneous stimulation occurs when the patient can identify the area of stimulation on the involved extremity when touched alone but cannot identify the area of stimulation on the involved extremity when simultaneously touched on the same region on both extremities. This occurs because the neurons that carry tactile sensation from the involved side are cortically overridden by the neurons carrying sensation from the uninvolved side.

5. *Pain.* Pain scales (see Figures 4.9–4.11) are used to document patient self-reports of pain intensity.

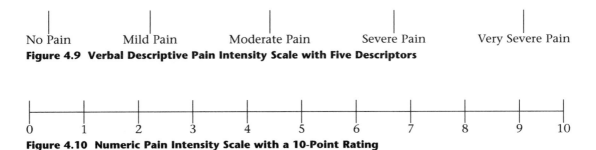

No Pain   Mild Pain   Moderate Pain   Severe Pain   Very Severe Pain

Figure 4.9 Verbal Descriptive Pain Intensity Scale with Five Descriptors

0   1   2   3   4   5   6   7   8   9   10

Figure 4.10 Numeric Pain Intensity Scale with a 10-Point Rating

No Pain   Pain as Bad as It Could Possibly Be

Figure 4.11 Visual Analogue Scale (Cline, Herman, Shaw, & Morton, 1992)

RED FLAGS
SIGNS AND SYMPTOMS OF SENSORY IMPAIRMENT

A red flag is an indicator or symptom of dysfunction. If a patient is observed displaying any of the following red flags, it is likely that the patient possesses sensory impairment.

SPECIAL SENSORY IMPAIRMENT

Olfaction:

- Anosmia, or loss of smell
- Inability or difficulty distinguishing between different smells
- Inability or difficulty identifying spoiled foods by smell.

Vision:

- Complaints of blurry or fuzzy print
- Difficulty reading print
- Complaints of lighting issues
- Shifting of page position
- Observed viewing out of the corner of eye
- Difficulty reading signs
- Poor navigational skills
- Abbreviated scanning
- Bumping into objects
- Complaints of image and objects darker on one side.

Gustation:

- Decreased or lost taste sensation on anterior tongue for sweet taste
- Decreased or lost taste sensation on posterior tongue for bitter taste
- Inability or difficulty identifying spoiled food by taste.

Hearing, balance, and equilibrium:

- Decreased or absent hearing
- Missing pieces of conversation
- Decreased balance
- Decreased or impaired ambulation
- Increased falls
- Complaints of vertigo.

PRIMARY SOMATOSENSORY IMPAIRMENT

Light touch:

- Difficulty with fine motor tasks
- Difficulty or inability manipulating fasteners or coins
- Decreased awareness of body segment being touched
- Difficulty shaving or applying make-up.

Pain:

- Burning or aching pain sensation
- Inability concentrating secondary to pain
- Sleep difficulty secondary to pain
- Recurrent self-injury secondary to decreased sensation
- Avoidance of specific self-care tasks secondary to infliction of pain.

Temperature:

- Recurrent burns
- Difficulty or inability distinguishing temperature changes on the skin surface
- Difficulty or inability responding to extreme temperature changes.

Proprioception:

- Difficulty with purposeful movement
- Difficulty with object manipulation
- Noted or increased clumsiness during functional upper-extremity reach
- Poor fine motor control
- Poor gross motor control
- Difficulty with transitional movements.

Vibration:

- Difficulty or inability detecting vibration.

Tactile localization:

- Difficulty with object manipulation
- Awkwardness and clumsiness during movements
- Difficulty or inability scratching an itch.

CORTICAL OR SECONDARY SOMATOSENSORY IMPAIRMENT

Two-point discrimination:

- Poor regulation of grip
- Difficulty with writing
- Dropping utensils during grooming and eating.

Astereognosis:

- Difficulty or inability identifying objects when vision is occluded
- Difficulty or inability identifying objects in a pants pocket
- Difficulty or inability identifying items in a grocery bag
- Difficulty or inability finding keys in a pocketbook.

Ahylognosia:

- Difficulty using textures in ADL tasks
- Difficulty manipulating hook-and-loop fasteners
- Difficulty discriminating between coins and other round, smooth objects in one's pockets (e.g., buttons, subway tokens)
- Difficulty discriminating between clothing materials in a dark closet (selecting the silk blouse instead of the cotton one).

Amorphagnosia:

- Difficulty using shapes in ADL tasks
- Difficulty discriminating between coin sizes
- Difficulty selecting the correct-sized measuring cup in a meal preparation task
- Difficulty discriminating packets of sugar, salt, pepper, and artificial sweetener at a restaurant
- Difficulty discriminating self-care items for morning grooming (e.g., soap pump dispenser, toothpaste, tube of body lotion, hairbrush, toothbrush, shaver).

Agraphesthesia:

- Inability to identify written letters on one's hand.

Extinction of simultaneous stimulation:

- Difficulty with bimanual tasks
- Dropping items when held in two hands
- Difficulty with bilateral coordination
- Difficulty manipulating a knife and fork.

AVAILABLE IN-DEPTH ASSESSMENTS

If sensory screening detects neurologic impairment, the following in-depth assessments are available:

Activities-Specific Balance Confidence Scale (ABC)
L. E. Powell, A. M. Myers
Adults with balance deficits
Powell, L. E., & Myers, A. M. (1995). The Activities-Specific Balance Confidence (ABC) Scale. *Journal of Gerontology, 50A*(1), M28–M34.

Berg Balance Scale (Berg)
K. O. Berg, S. L. Wood-Dauphinee, J. I. Williams, B. Maki
Adults with balance deficits
Berg, K. O., Wood-Dauphinee, S. L., Williams, J. I., & Maki, B. (1992). Measuring balance in the elderly: Validation of an instrument. *Canadian Journal of Public Health, 83,* S7–S11.

Brain Injury Visual Assessment Battery for Adults
M. Warren
Adults with visual deficits secondary to neurologic impairment
Precision Vision
745 North Harvard Ave.
Villa Park, IL 60181

Central Visual Field Testing for Low-Vision Patients
D. C. Fletcher
Adults with visual deficits
Mattingly International, Inc.
938-K Andreasen Dr.
Escondido, CA 92029

Clinical Test of Sensory Interaction in Balance (CTSIB)
M. Watson
Adults with balance deficits
Watson, M. (1992). Clinical Test of Sensory Interaction in Balance. *Physiotherapy Theory and Practice, 8,* 176–178.

Comprehensive Test of Visual Functioning (CTVF)
S. L. Larson, E. Buethee, G. J. Vitali
Children, adolescents, and adults with neurologic impairment
Slosson Educational Publications, Inc.
PO Box 280
East Aurora, NY 14052-0280

Falls Efficacy Scale (FES)
M. E. Tinetti, D. Richman, L. Powell
Elderly persons with balance deficits
Tinetti, M. E., Richman, D., & Powell, L. (1990). Falls efficacy as a measure of fear of falling. *Journal of Gerontology, 45,* P239–P243.

Functional Reach Test (FRT)
P. W. Duncan, D. K. Weiner, J. Chandler, S. A. Studenski
Adults with balance deficits secondary to neurologic damage
Duncan, P. W., Weiner, D. K., Chandler, J., & Studenski, S. A. (1990). Functional reach: A new clinical measure of balance. *Journal of Gerontology, 45,* 192–197.

Jebsen Hand Function Test
R. H. Jebsen, N. Taylor
Adults with neurological impairment
Sammons Preston Inc.
PO Box 50710
Bolingbrook, IL 60440-5071
(800) 323-5547

McGill Pain Questionnaire
R. Melzack
Patients with pain
Melzack, R. (1983). *Pain measurement and assessment.* New York: Raven.

MN Acuity Chart
Adults with visual impairments
Lighthouse Low Vision Products
36-02 Northern Blvd.
Long Island City, NY 11101

Morse Fall Scale (MFS)
J. M. Morse
Adults with balance deficits
Janice M. Morse
Pennsylvania State University
School of Nursing
201 Health Human Development East
University Park, PA 16802-6508

Nine-Hole Peg Test
V. Mathiowetz, K. Weber, N. Kashman,
G. Volland
Adults with fine motor coordination deficits
secondary to neurologic and/or musculoskeletal
impairment
Mathiowetz, V., Weber, K., Kashman, N., &
Volland, G. (1985). Adult norms for the Nine-
Hole Peg Test of finger dexterity. *Occupational*
Therapy Journal of Research, 5, 24–38.

O'Connor Finger and Tweezer Dexterity Tests
J. O'Connor
Adults with fine motor coordination deficits secondary
to neurologic and/or musculoskeletal impairment
Sammons Preston Inc.
PO Box 5071
Bolingbrook, IL 60440-5071
(800) 323-5547

Pain Apperception Test (PAT)
D. V. Petrovich
Adults with pain
Western Psychological Services
12031 Wilshire Blvd.
Los Angeles, CA 90025

Pepper Test (VSRT)
G. R. Watson, S. Whittaker, M. Steciw
Adolescents and adults with visual impairment
Mattingly International, Inc.
938-K Andreason Dr.
Escondido, CA 92029

Purdue Peg Board
J. Tiffin
Adults with fine motor deficits
Lafayette Instrument Company
3700 Sagamore Pkwy. North

PO Box 5729
Lafayette, IN 47903

Semmes–Weinstein Monofilaments
J. Semmes, M. Weinstein
Adults with sensory problems
Sammons Preston Inc.
PO Box 5071
Bolingbrook, IL 60440-5071
(800) 323-5547

Revised Sheridan Gardiner Test of Visual Acuity
M. D. Sheridan, P. Gardiner
Age 5 and over
Keeler Instruments, Inc.
456 Parkway
Broomall, PA 19008

Timed Get Up and Go
D. Podsiadlo, S. Richardson
Adults with balance deficits
Podsiadlo, D., & Richardson, S. (1991). The
Timed Get Up and Go: A test of basic functional
mobility for frail elderly persons. *Journal of the*
American Geriatric Society, 39, 142–148.

Tinnetti Balance Test of the Performance-Oriented Assessment of Mobility Problems (Tinnetti)
M. E. Tinnetti
Adults with balance deficits
Tinnetti, M. E. (1986). Performance oriented
assessment of mobility in elderly patients. *Journal*
of the American Geriatric Society, 34, 119–126.

For further information about these assessments,
refer to the following resources:

Asher, I. E. (1996). *Occupational therapy*
assessment tools: An annotated index (2nd ed.).
Bethesda, MD: American Occupational
Therapy Association.
Impara, J. C., Plake, B. S., & Murphy L. L. (Eds.).
(1998). *The thirteenth mental measurements*
yearbook. Lincoln: University of Nebraska
Press.

Murphy, L. L., Impara, J. C., & Plake, B. S. (Eds.). (1999). *Tests in print: An index to tests, test reviews, and the literature on specific tests* (Vols. 1 & 2). Lincoln: University of Nebraska Press.

Plake, B. S., Impara, J. C., & Murphy, L. L. (Eds.). (1999). *The supplement to* The Thirteenth Mental Measurements Yearbook. Lincoln: University of Nebraska Press.

For further information about the Warren Near Text Card, the Lea Symbols Near Vision Test Card, and the Lighthouse Near Acuity Text Card, contact Mattingly International, Inc., 938-K Andreason Dr., Escondido, CA 92029.

Reference

Cline, M. E., Herman, J., Shaw, E. R., & Morton, R. D. (1992). Standardization of the Visual Analogue Scale. *Nursing Research, 41,* 378–380.

SENSORY SCREENING

Patient Name _____   Date _____

Sensation	No Impairment Detected	Impairment Detected	Body Region
Olfaction			
Vision			
Near distance			
Far distance			
Taste			
Sweet			
Bitter			
Sour			
Salty			
Hearing			
Balance			
Equilibrium			
Light touch			
Pain			
Temperature			
Proprioception			
Tactile localization			
Vibration			
Two-point discrimination			
Stereognosis			
Ahylognosia			
Amorphagnosia			
Graphesthesia			
Extinction of simultaneous stimulation			

Motor
SCREENING

FUNCTIONAL IMPLICATIONS OF MOTOR IMPAIRMENT

Motor control depends on an interrelationship between the central nervous system (CNS) and the peripheral nervous system, which work together systematically to provide us with the ability to move freely with precise, coordinated movements. *Movements begin reflexively* in nature. As infants begin to interact and explore their environment, new movement patterns are performed and stored on a cortical level. As our CNS matures, primitive reflexive patterns are integrated and learned. Movement patterns such as walking, reaching for a glass, and rolling in bed are integrated subcortically and become automatic. Previously and newly learned movement patterns reflect the integration and maturation of our CNS, allowing us to perform a variety of movement patterns that are now accessed on a subconscious level, creating complex motor plans.

Normal movement relies on constant communication between the sensory and motor systems. For example, a patient reaching behind him- or herself to place an arm in a shirtsleeve requires sensory information to guide that particular motor movement, making motor correction if necessary. Our intent to move, whether on a voluntary or unconscious level, is initiated in the *primary motor area*—also referred to as the *precentral gyrus.* This cortical area is the origin of the *corticospinal tracts,* which carry descending motor messages down through the *spinal cord.* Before reaching the spinal cord, descending motor messages travel through the *internal capsule* to the *thalamus,* where sensory and motor information is relayed and processed.

The *basal ganglia* operate on a subcortical level and have several functions contributing to motor control:

- Integration and storage of stereotypic movements such as swimming, riding a bike, or filling up a glass of water, once–cortically learned movements that become unconscious, automatic motor patterns enabling a patient to get dressed, ambulate to the bathroom, and bring a spoon to his or her mouth—all with little conscious cortical input

- Influence of tone and posture

- Initiation and disinhibition of movement

- Error correction during movement.

For example, a patient with a basal ganglia lesion may present with increased upper-extremity and trunk tone, inhibiting a patient's ability to sit adequately and safely in a wheelchair.

Preservative motor movements during grooming tasks may prevent the patient from performing the task independently or in a timely manner. Dressing tasks that involve manipulation of fasteners may become tedious and frustrating due to an inability to refine precision movements.

The *brain stem* receives descending motor input from the cortex. Its functions include

- Mediating tonal patterns of the trunk and limbs in relation to head position, and head position in relation to body position

- Balance correction

- Fine-tuning of motor movements

- Integration of vestibular input.

For example, the brain stem allows an individual to perform bed mobility, functional transfers, rotational and flexion movements during dressing while maintaining normal balance, and upper-extremity movement necessary to perform activities of daily living (ADL) tasks.

The *descending motor tracts* belonging to the pyramidal system include the *lateral* and *anterior corticospinal tracts* and the *corticobulbar tract*. The corticospinal tract is responsible for carrying conscious/voluntary motor information from the primary motor cortex to the skeletal muscles. The idea of quenching one's thirst is initiated in the primary motor cortex and brought to fruition when the skeletal muscles are able to carry out the desired action. The corticobulbar tract is responsible for projection of motor information, having a motor component, to the cranial nerve nuclei:

- Cranial nerves 3, 4, and 6 are the control centers for extraocular eye movements for visual scanning, binocular vision, and reading.

- Cranial nerves 5 and 7 are the control centers for the facial muscles used for verbal and nonverbal language.

- Cranial nerves 9, 10, 11, and 12 are the control centers for swallowing, eating, and speaking.

Extrapyramidal descending motor fibers include the following:

- *Medial longitudinal fasciculus* allows for coordination of head and neck movements. For example, this allows a patient to visually scan for items successfully when shopping in a store because he or she can coordinate extraocular movements with head and neck movement to increase the field of visual scanning.

- *Vestibulospinal tract* is responsible for the muscles that control posture and stance through the facilitation of extensor tone against gravity. A patient is able to activate appropriate extensor muscles when ambulating upright.

- *Rubrospinal tract* facilitates extensor tone primarily within the lower extremities. Upper-extremity extensor muscles are required to allow for forward reach when donning socks, and lower-extremity extensor muscles are required when transitioning from sitting to standing position.

- *Medullary reticulospinal tract* facilitates flexor tone and helps regulate cardiovascular responses such as heart rate, blood pressure, and respiration.

- *Pontine reticulospinal tract* facilitates antigravity muscles—extensor tone. For example, patients can transition from supine position to sitting position in preparation for functional transfers or maintain their heads in an upright position for watching television or reaching.

The *cerebellum* receives information through the *spinocerebellar tracts*. The cerebellum functions to coordinate gross and fine motor skills (required for eating or buttoning a shirt), regulation of equilibrium (enabling one to make a bed while maintaining balance), and muscle tone (enabling a person to fill a glass of water without dropping the glass). Sensory information, or proprioceptive information, is received by the cerebellum. Proprioceptive information allows an individual to make the appropriate motor responses for precise, smooth, coordinated movement and balance reactions. These are performed on an unconscious level. The cerebellum has three lobes that contribute to normal movement:

- *Flocculonodular lobe* controls balance, trunk control, and postural reflexes. For example, patients are able to shower safely while maintaining good dynamic balance and normal posture.

- *Anterior lobe* controls postural adjustment, extremity movement and control, and automatic movements necessary for static and dynamic ADL tasks.

- *Posterior lobe* controls motor planning and timing and also coordination of multiple muscle groups, allowing an individual to speak fluently or engage in a game of catch.

The *medulla* carries the descending motor messages to the spinal cord via motor spinal cord tracts. Motor messages travel down the spinal cord tracts and synapse with interneurons in the ventral horn. The interneurons synapse with motor spinal nerves before they exit the ventral horn through the ventral rootlet. The rootlet joins the ventral root, delivering the final motor message to the skeletal muscle for execution of the motor command. This is achieved in the peripheral nervous system.

Disease and trauma involving the CNS will have a negative impact on the motor control process and the ability to perform functional ADL and instrumental activities of daily living tasks. Patients with *cerebral lobe lesions* may exhibit *akinesia*, *apraxia*, and *paratonia*. Functionally, patients may demonstrate an inability to motorically initiate weight shifts or change sitting position when uncomfortable. Individuals may require supervision during functional ambulation due to an ataxic gait that decreases safety during outdoor mobility. Patients may exhibit limb disuse, presenting as though the limb is hemiplegic.

Patients with *frontal lobe lesions* may exhibit *apraxia* as evidenced by a patient's inability to use eating utensils correctly or to demonstrate appropriate motor actions when performing a morning grooming routine.

Patients with *upper motor lesions* may exhibit spasticity below the lesion level. For example, increased flexor tone of the trunk and upper extremity may present major obstacles because the patient is unable to use the trunk and arm to perform basic self-care tasks such as dressing, feeding, bathing, or grooming. *Flaccidity* is exhibited at the lesion level, as patients may be unable to voluntarily recruit muscles necessary for performing mobility and ADL tasks on the involved side.

Patients with lesions to the primary motor area present with *contralateral loss* of voluntary muscle movement, spasticity of the *distal musculature,* and *hyperreflexia.* Patients with stroke may exhibit increased flexor tone in the upper extremity, interfering with functional use of that extremity.

Lesions involving the *internal capsule* may exhibit *contralateral spastic paralysis* and hyperreflexia. Patients with these specific lesions may demonstrate an inability to use the involved extremity for normal gait during community tasks or in bimanual tasks such as unloading a shopping cart or folding laundry.

Patients with brain stem lesions *above the level of decussation* may exhibit contralateral spastic paralysis, impeding the ability to perform all self-care tasks. Lesions *below the level of decussation* may show *ipsilateral loss* of voluntary motor control, spasticity in the distal musculature below the lesion level, flaccidity at the lesion level, and hyperreflexia and *decerebrate rigidity.*

Brain stem lesions of the *pons* present *ocular abnormalities.* For example, patients may present with strabismus and diplopia, impeding reading, negotiation of steps, and reaching. *Quadriplegia* and *facial weakness* may occur in *locked-in syndrome,* and *pure motor weakness* may be observed in *lacunar infarcts.*

Brain stem lesions affecting the medulla may impede a patient's balance secondary to complaints of vertigo, or *ipsilateral ataxia* may occur, affecting a person's ability to reach into the refrigerator and pour a glass of water without significant difficulty or spillage.

Pathology to the basal ganglia present

- Difficulty initiating and stopping movement as observed in the increased time required for a patient to dress for therapy

- Increased involuntary extraneous movements such as lip smacking and writhing movements that impede an individual's social confidence

- Hypertonicity and rigidity, making functional mobility and ADL performance difficult and dangerous.

Patients with lesions to the cerebellum clinically present with ataxic gait, trunk and upper extremity hyperreflexia, intention tremors, dysmetria, disturbances in extensor tone, nystagmus, and equilibrium and propriocetive disorders. Functionally, patients may exhibit deficits in feeding or applying make-up secondary to ataxic movement patterns. Turning newspaper or book pages becomes arduous, secondary to tremors and asymmetric movements. Kitchen navigation for meal preparation or the ability to navigate a grocery store becomes tiresome and dangerous due to unsteady gait. Deficits in reading due to poor line and page navigation are frustrating, secondary to nystagmus.

Lesions to the descending motor tracts present with deficits in tone and postural reflexes, leading to difficulty performing mobility and ADL tasks. Patients with lesions to the corticobulbar tracts may present with oculomotor muscle involvement, impeding visual scanning and binocular vision. Areas such as reading, writing, step negotiation, and shaving are affected secondary to diplopia. Weakness in facial muscles contributes to food pocketing, drooling, and slurred speech. Patients are at increased risk for aspiration pneumonia. Patients with lesions to the corticospinal tracts may demonstrate difficulty with fine coordination of limb use, such as placing a pierced earring into the lobe and manipulating one's foot into a shoe.

Lesions to the vestibulospinal tract may result in deficits in maintaining adequate activation of trunk extensors while standing at the sink performing oral and facial hygiene or regaining erect trunk posture when reaching down for a dropped item. Lesions to the rubrospinal tract may result in deficits in the ability to anteriorly shift weight in preparation for transfers, or the ability to wash one's face due to an overexcitation of the extensor musculature. Patients with lesions to the medullary reticulospinal tract functionally present with difficulty in lower body dressing and forward reach, as movement is dominated by extensor tone. These patients may exhibit altered blood pressure, heart rate, and respiration, which must be closely monitored in therapy. Patients with lesions to the pontine reticulospinal tract may present with significant difficulty performing all ADL tasks secondary to rigidity. Patients with medial longitudinal fasciculus lesions may present with challenges in wheelchair positioning, poor visual scanning, and decreased communication abilities secondary to poor coordination of head and neck control.

MOTOR SCREENING PROCEDURES

Motor screenings should be performed on all individuals with traumatic brain injury, acquired brain injury, CNS lesions, and neuromuscular deficits. It is important to conduct a thorough chart review because vital information can be ascertained:

- *Diagnosis.* For example, with patients with cerebellar cerebrovascular accident, one might anticipate deficits in coordination, balance, and equilibrium reactions, hence affecting the ability to bring a spoon to their mouth or maintain their balance during showering.

- *Lesion/injury.* For example, if a patient sustained a parietal lobe lesion on the dominant hemisphere, one might anticipate positive signs of ideomotor apraxia.

- *Medical history.* Has the patient had a history of another previously acquired brain injury or secondary neurological diagnosis that might affect existing movement?

A therapist would want to monitor blood pressure during treatment if the stroke etiology is hypertensive in nature.

- *Medication.* Knowledge regarding medications presently taken by patients is useful because some may cause tremors, changes in tone, or cognitive status, affecting the motor movements required to perform various mobility and self-care tasks.

- *Other discipline notes.* Nursing may report difficulty positioning the patient in a wheelchair before therapy secondary to fluctuating changes in tone throughout the day.

Knowledge of the patient's cognitive, hearing, and visual acuity status is necessary for test administration. Refer to Section 1, "Cognitive Screening," for orientation, arousal, recognition, attention, memory, comprehension, perception, and the ability to follow directions. The therapist should use clinical judgment regarding the use of simple and complex commands when administering testing. Complex tasks can be broken down into simpler components to ensure successful accurate testing. A thorough sensory screening should be administered before the motor screening, as motor movement depends on the sensory feedback loop. Refer to Section 4, "Sensory Screening."

It is necessary to establish the patient's understanding of directions and ability to follow commands as well as to identify an alternate response system if the patient is not verbal. Always explain the goals and purpose of the testing procedure to the patient and elicit the patient's feedback to ensure that instructions are understood. The therapist should emphasize cooperation from the patient, as it is necessary to ensure the most accurate result.

The testing environment should be in a quiet room with minimal distractions and maximal privacy. The patient should be comfortable and relaxed. Access to furniture to create a variety of patient positions such as a bed, a chair, and a stable standing surface is ideal to posturally challenge the patient, as they may perform tasks significantly different in supine than in sitting positions. The patient should wear loose-fitting clothing or a gown because it is important for the therapist to visually evaluate symmetry and movement throughout the trunk, upper extremities, and lower extremities. The therapist should provide the patient with a visual demonstration of each movement evaluated to ensure maximal patient understanding. Observation of patient performance, transitional movements, and voluntary and involuntary movements and postures should begin immediately.

Evaluation should begin on the nonaffected side when testing range of motion, tone, and functional muscle strength. Goniometric measurements are indicated only when orthopedic limitations accompany neurological diagnosis or when the therapist is unable to make an accurate estimation of range of motion. When performing functional mobility, transfers, or transitional movements, always check the patient's ambulation status, guarding, and appropriate assistive device used, if any. The therapist should carefully guard the patient to eliminate risk of falls or injury.

Motor screening is an ongoing evaluation because the patient's movement, tone, and reflex patterns change and should be evaluated at various times of the day. Record the test results on the form at the end of this section.

OBSERVATION

1. *Bed/supine.*

- Does the patient's position look relaxed?

- Is there any posturing noted?

- Can the patient move around freely?

- Can the patient roll, transition from supine to sit, and scoot in bed?

- What does the quality of the patient's movement look like?

- Can the patient move all extremities?

- Is there increased tone noted in the limbs or trunk?

- Can the patient's head move independently of his or her trunk?

- Do movements requiring great effort increase the patient's tone?

2. *Sitting.*

- Can the patient support him- or herself in supported sitting or unsupported sitting?

- Does the patient have head control?

- Can the patient move his or her head independently of the body?

- Can the patient move his or her body independently of the head?

- Can the patient maintain trunk control while engaging the upper extremities in an activity?

- Can the patient isolate the upper trunk from the lower trunk? Can he or she use the upper extremities for activity?

- Do stressful movements increase tone (see Figure 5.1)?

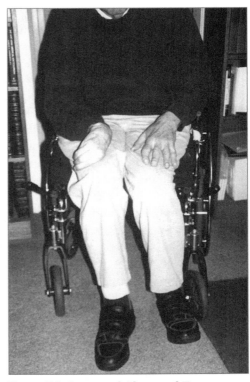

Figure 5.1 Increased Abnormal Tone Elicited During Sitting

- What is the quality of the patient's movement during basic self-care tasks?

- Are tremors noted during reach? What is the quality of limb use during ADL tasks?

3. *Standing.*

- What is the quality of the patient's movement as he or she transitions into standing (see Figure 5.2)?

- Does movement require much effort?

- Does the patient move synergistically?

- Are both sides of the body working together?

- Is the patient's weight evenly distributed throughout the body?

- Can the patient perform weight shifts during upper-extremity use?

4. *Functional ambulation.*

- Are tonal patterns evident during functional ambulation (see Figure 5.3)?

- Does the quality of movement change with fatigue or stress?

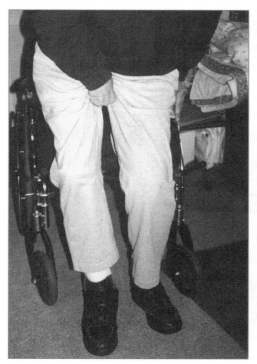

Figure 5.2 Increased Abnormal Tone Elicited During Sit-to-Stand

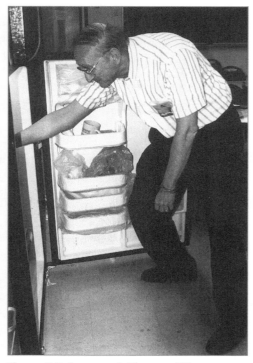

Figure 5.3 Increased Abnormal Tone Elicited During Reaching

- Can the patient perform multistep tasks with good motor control?

- Are extraneous movements observed during ambulation?

Note: If yes is indicated in any of the above questions, additional motor screening should be performed. Patients should be screened in multistimulation environments at various times of the day while performing ADL routines and during strenuous activities. Therapists should note the smoothness of the patient's movements (see Table 5.1), incorporation of accessory or secondary muscles to perform a task, tonal patterns (see Table 5.2), and the amount of time required to perform the movement.

SCREENING FOR AVAILABLE MOVEMENT

1. *Active range of motion (AROM)* is the amount of joint motion achieved during unassisted voluntary joint motion.

Table 5.1 Manual Muscle Testing Grading Scale

Grade	Description
0	No muscle contraction detected
Trace	No joint motion, but contraction felt when muscle palpated
Poor−	Less than full ROM in gravity-eliminated position
Poor	Full ROM in gravity-eliminated position
Poor+	Full ROM in gravity-eliminated position and able to withstand minimal resistance
Fair−	Less than full ROM against gravity
Fair	Full ROM against gravity
Fair+	Full ROM against gravity and able to withstand minimal resistance
Good−	Full ROM against gravity and able to withstand less than moderate resistance
Good	Full ROM against gravity and able to withstand moderate resistance
Normal	Full ROM against gravity and able to withstand maximum resistance

Note. ROM = range of motion. From "Muscle Group Testing," by S. Brunnstrom, 1941, *Physiotherapy Review, 21*, pp. 3–21. Copyright 1941 by the American Physical Therapy Association. Reprinted with permission.

Table 5.2 Modified Ashworth Scale

Grade	Description
0	No increase in muscle tone
1	Slight increase in tone manifested by a catch and release or by minimal resistance at the end of the range of motion (ROM) when the affected part is flexed or extended
−1	Slight increase in muscle tone manifested by a catch followed by minimal resistance throughout the remainder (less than half) of the ROM
2	More marked increase in muscle tone through most of the ROM, but affected part is easily moved
3	Considerable increase in muscle tone, making passive movement difficult
4	Affected part rigid in flexion or extension

Note. From "Interrater Reliability of a Modified Ashworth Scale of Muscle Spasticity," by R. W. Bohannon and R. B. Smith, 1987, *Physical Therapy, 67*, p. 207. Copyright 1987 by the American Physical Therapy Association. Reprinted with permission.

Clinical screening for AROM:

- Place the patient in a seated position. Postural support is indicated if trunk balance is less than fair.

- Ask the patient to look over his or her right shoulder, over the left shoulder (cervical rotation), up at the ceiling, and down at the floor (cervical flexion/extension).

- Ask the patient to move his or her right ear to the right shoulder. Repeat on the left side (lateral cervical flexion).

- Ask the patient to shrug his or her shoulders up toward the ears and release (shoulder elevation).

- Ask the patient to squeeze his or her shoulder blades back as though they are touching (shoulder retraction).

- Ask the patient to place his or her right arm straight out in front and then raise it toward the ceiling as though he or she is reaching for an item on the top shelf. Repeat on the left side (shoulder flexion).

- Ask the patient to raise his or her right arm out to the side and then as high above the head as possible. Repeat on the left side (shoulder abduction).

- Ask the patient to place his or her hands behind the head with elbows opened wide as though he or she is braiding hair (shoulder external rotation).

- Ask the patient to place the hands behind the back as though she is going to fasten a bra or he is tucking a shirt into pants (shoulder internal rotation).

- Ask the patient to touch his or her shoulders and extend the arms straight out (elbow flexion/extension).

- Ask the patient to hold his or her arms out in front and face the palms upward toward the ceiling and then turn them downward toward the floor (supination/pronation).

- Ask the patient to bend his or her right wrist upward with fingers pointing toward the ceiling, or ask the patient to make a stop gesture with his or her hand. Repeat on the left side (wrist extension).

- Ask the patient to bend his or her right wrist downward with fingers pointing toward the floor, or ask the patient to make the wrist go limp. Repeat on left side (wrist flexion).

- Ask the patient to move his or her right wrist radially (with fingers pointed inward) and ulnarly (with fingers pointed outward). Repeat on the left side (see Figures 5.4 and 5.5).

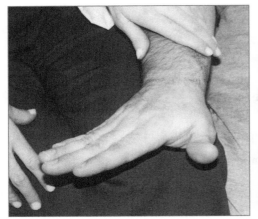

Figure 5.4 Ulnar Deviation

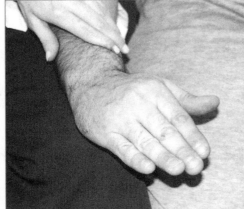

Figure 5.5 Radial Deviation

■ Ask the patient to open and close his or her right hand fully. Repeat on the left side (finger flexion/extension and finger abduction/adduction).

■ Ask the patient to march in place while sitting (hip flexion).

■ Ask the patient to bend and then extend the right knee straight out in front. Repeat on the left side (knee flexion/extension).

■ Ask the patient to make a circle with the right ankle. Repeat on the left side (ankle inversion/eversion and dorsiflexion/plantar flexion).

Functional screening for AROM:

■ Place an object behind the patient and have him or her look for it and retrieve the object while sitting and then standing.

■ Have the patient reach overhead for an object.

■ Have the patient don a button-up shirt or blouse.

■ Have the patient pick items up off of the floor.

■ Observe the patient comb his or her hair.

■ Have the patient cross his or her legs to don a pair of pants.

☞ A range of motion impairment is indicated if the patient is fewer than 10° from the measurements on the norm chart (see Table 5.3) or the patient does not have the available range of motion to perform the movement of the functional task.

Note: The therapist should observe the quality of movement, timing, and ability to control the limbs and trunk.

Table 5.3 Functional Joint Range of Motion

Functional Movement	Functional ROM in Degrees	Functional Movement	Functional ROM in Degrees
Cervical flexion	0–60	Index finger abduction	No norm
Cervical extension	0–70	Index finger adduction	No norm
Shoulder flexion	0–180	Middle finger MP flexion	0–90
Shoulder extension	0–60	Middle finger MP hyperextension	0–45
Horizontal abduction	0–90	Middle finger PIP flexion–extension	0–100
Horizontal adduction	0–45	Middle finger DIP flexion–extension	0–90
Shoulder abduction	0–180	Ring finger MP flexion	0–90
Internal rotation	0–80	Ring finger MP hyperextension	0–45
External rotation	0–60	Ring finger PIP flexion–extension	0–100
Elbow flexion–extension	0–150	Ring finger DIP flexion–extension	0–90
Supination	0–80	Little finger MP flexion	0–90
Pronation	0–80	Little finger MP hyperextension	0–45
Wrist flexion	0–80	Little finger PIP flexion–extension	0–100
Wrist extension	0–70	Little finger DIP flexion–extension	0–90
Ulnar deviation	0–30	Hip flexion	0–120
Radial deviation	0–20	Hip extension	0–30
Thumb CM flexion	0–15	Hip abduction	0–45
Thumb CM extension	0–20	Hip adduction	0–30
Thumb MP flexion–extension	0–50	Hip internal rotation	0–45
Thumb IP flexion–extension	0–80	Hip external rotation	0–45
Thumb abduction	0–70	Knee flexion–extension	0–135
Thumb opposition	No norm	Ankle dorsiflexion	0–20
Index finger MP flexion	0–90	Ankle plantar flexion	0–50
Index finger MP hyperextension	0–45	Ankle inversion	0–35
Index finger PIP flexion–extension	0–100	Ankle eversion	0–15
Index finger DIP flexion–extension	0–90		

Note. ROM = range of motion. From "Goniometry: The Measurement of Joint Motion," by F. H. Krusen and F. J. Kottke, 1971, *Handbook of Physical Medicine and Rehabilitation, 2nd ed.,* pp. 45–56. Copyright 1971 by Elsevier Science. Reprinted with permission from Elsevier Science.

2. *Passive range of motion (PROM)* is the amount of available joint motion when the therapist moves the joint through full range of motion. *Note:* Use of a goniometer is indicated when orthopedic deficits accompany the neurological deficit. PROM should be estimated for the approximate degree measurement or available range. For example, shoulder flexion is approximately 120° or one-half to three-quarters range of motion.

The therapist should assess the noninvolved extremity first. If a limitation is present, notation should be documented both quantitatively (measured numbers) and qualitatively (narrative description), for example, (a) hard end-feel in which movement can no longer occur due to a hard bony feel at the end of the joint range and (b) spongy or soft end-feel in which movement has a springlike feel at the joint end range. Does the patient exhibit pain? Does the muscle feel as though it catches or stops through the available range?

Clinical screening for PROM:

- Place the patient in a seated position. Postural support is indicated if trunk balance is less than fair.

- Turn the patient's head to look over his or her right shoulder, over left shoulder (cervical rotation), up at ceiling, and down at floor (cervical flexion and extension).

- Laterally flex the patient's head so that the right ear moves toward the right shoulder. Repeat on left side.

- Mobilize the patient's shoulders up toward the ears and release (shoulder elevation).

- Mobilize the patient's shoulder blades backward (shoulder retraction).

- Mobilize the patient's shoulder blades forward (shoulder protraction).

- Mobilize the patient's arm so that it is straight out in front and then lifted high above the head (shoulder flexion).

- Mobilize the patient's arm so that it is straight out to the side and then lifted above the head (shoulder abduction).

- Mobilize the patient's arms so that they are reaching behind the back as if to tuck a shirt into the pants (shoulder internal rotation).

- Mobilize the patient's arms so that they are reaching behind the head as if to braid hair (shoulder external rotation).

- Mobilize the patient's elbow so that it is straight (elbow extension).

- Mobilize the patient's elbow so that the elbow is bent and fingers are reaching to touch the shoulder (elbow flexion).

- Mobilize the patient's forearm so that the palms are face up toward the ceiling (forearm supination).

- Mobilize the patient's forearm so that the palms are face down toward the floor (forearm pronation).

- Mobilize the patient's wrist so that it is bent with fingers pointing down toward the floor (wrist flexion).

- Mobilize the patient's wrist so that it is bent with fingers pointing up toward the ceiling (wrist extension).

- Mobilize the patient's wrist so that it is bent with fingers deviated outward (ulnar deviation) (see Figure 5.4).

- Mobilize the patient's wrist so that it is bent with fingers deviated inward (radial deviation) (see Figure 5.5).

- Mobilize the patient's fingers so that they are fully spread apart and held straight out (finger abduction).

- Mobilize the patient's fingers so that they are tightly touching and held straight out (finger adduction).

- Mobilize the patient's fingers into a fist (finger flexion).

- Extend the patient's fingers straight out (finger extension).

- Bend and straighten the patient's hip (hip flexion/extension).

- Bend and straighten the patient's knees (knee flexion/extension).

- Mobilize the patient's ankles in a circular motion.

- Mobilize the patient's ankles so that the toes are pointed to the floor (plantar flexion).

- Mobilize the patient's ankles so that the toes are pointed toward the ceiling (dorsiflexion).

Functional screening for PROM:

- Observe the patient while performing assisted bed mobility; transfers to the wheelchair, commode, and chair; dressing; bathing; and overhead reaching.

☛ Passive range of motion restrictions are indicated if the available degree of motion is 10° fewer than the chart norms (see Table 5.3) or if the restriction inhibits a patient functionally during mobility or ADL tasks.

3. *Tone* is defined as resistance of a muscle to stretch. *Note:* The therapist should encourage the patient to allow the joint and limb to be moved, stating, "Try not to resist the movement." Care should be taken when performing a tonal assessment of the cervical area because injury may occur if too much pressure is applied to the cervical joints. Each joint segment should be moved four times quickly within the allowable motion.

Clinical screening of tone:

- Place the patient in a comfortable position.

- Place open palms on each side of the patient's jaw. Quickly turn the patient's head from neutral position to the right. Repeat to the left side (see Figure 5.6).

- Place hands on top of the patient's head. Quickly move the patient's head from neutral position into cervical flexion. Repeat for motion into extension (see Figure 5.7).

- Place hands along the patient's wrist and elbow. Quickly move the patient's shoulder from neutral position (extension) into shoulder flexion (see Figure 5.8). Repeat for extension. Repeat for abduction and adduction (see Figure 5.9).

- Place hands along the patient's wrist and posterior portion of the triceps. (Starting position of the limb is in 90° of abduction and 90° of elbow flexion.) Quickly move the patient's limb between starting position and internal rotation (see Figure 5.10). Repeat for external rotation.

Figure 5.6 Screening Tone: Quick Stretch into Cervical Lateral Flexion

Figure 5.7 Screening Tone: Quick Stretch into Cervical Flexion

Note: If the patient is unable to achieve 90° of abduction secondary to soft tissue restrictions and/or pain, place the limb into adduction and 90° of elbow flexion. Quickly move the forearm between external rotation and the starting position. Repeat for left side.

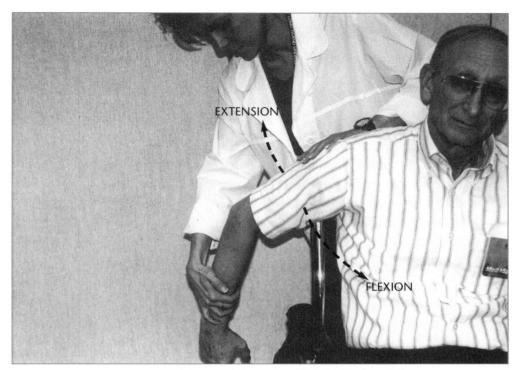

Figure 5.8 Screening Tone: Quick Stretch into Shoulder Flexion/Extension

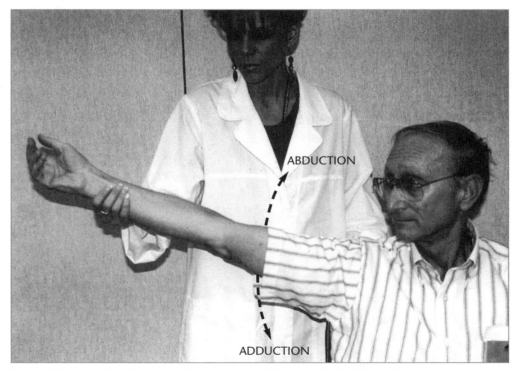

Figure 5.9 Screening Tone: Quick Stretch into Shoulder Abduction/Adduction

■ Place hands supporting the patient's olecrenon and wrist, with elbow extension in the starting position. Quickly move the patient's elbow into flexion (see Figure 5.11). Repeat for extension.

■ Place one hand supporting the patient's olecrenon and the other hand in a "shake hands" position. (Precautions should be taken to prevent stimulation to the palmar surface.) Starting position is forearm in midposition. Quickly move the forearm between supination and midposition (see Figure 5.12). Repeat for pronation.

■ Support the patient's forearm on a table or supported surface, allowing the wrist to hang freely. Starting position should place wrist in neutral. Place contact on the heads of the 2nd and 5th metacarpals. Quickly move the wrist between extension and neutral (see Figure 5.13). Repeat for flexion.

Functional screening of tone:

■ Observe the patient during bed mobility, transfers, and transitional

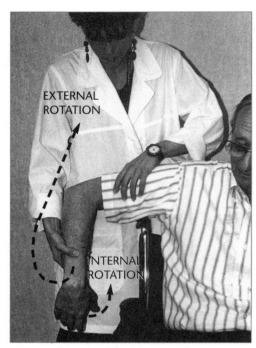

Figure 5.10 Screening Tone: Quick Stretch into Shoulder Internal/External Rotation

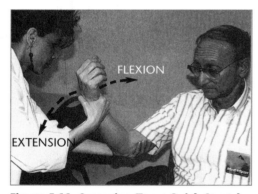

Figure 5.11 Screening Tone: Quick Stretch into Elbow Flexion/Extension

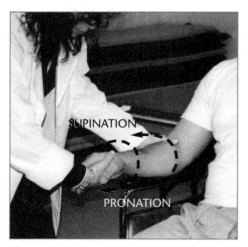

Figure 5.12 Screening Tone: Quick Stretch into Supination/Pronation

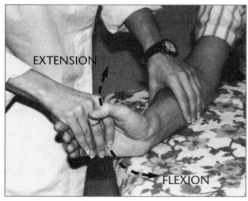

Figure 5.13 Screening Tone: Quick Stretch into Wrist Flexion/Extension

movements such as scooting, supine to sit, and sit to stand. For example, does upper-extremity tone increase when transitioning from supine to sit or sit to stand?

- Observe the patient during feeding, grooming, dressing, and functional ambulation. For example, can the patient use the involved extremity to assist with dressing? Can the extremity be used in bimanual tasks such as brushing one's teeth or cutting food?

- Observe the patient during various times of the day and in different environments (low and high stimulation). Are there changes in tone when the patient is fatigued later in the day or in the busy physical therapy gym?

- Observe the patient reaching in different positions, such as removal of shoes from the floor, reaching for a cup or the telephone, or cutting. Can the patient maintain forward trunk movements while eating?

☞ A tonal impairment is indicated if a "catch" or resistance is felt throughout the motion—referred to as *hypertonia*. Does the limb feel floppy or heavy, or is absence of movement noted (referred to as *hypotonia* or *flaccidity*)? "Beating or jumping throughout the motion" is referred to as *clonus*. Does the limb feel difficult to move in both directions (referred to as *rigidity*)? Does the patient demonstrate posturing of the limb or trunk or an inability to transition between movements (referred to as *dystonia*, or writhing/twisting movements)?

SCREENING FOR STRENGTH

Please note that MMT stands for manual muscle testing.

1. *Functional muscle strength (FMS)* is the amount of resistance a joint can sustain during a movement. *Note:* Do not assess FMS if increased tone is present, as testing may facilitate tone.

Clinical screening for FMS:

- Place the patient in a seated position.

- Ask the patient to shrug his or her shoulders toward the ears (shoulder elevation).

- Tell the patient, "Don't let me push your shoulders down." Apply resistance with palms toward the floor. Repeat on the involved side (see Figure 5.14).

- Ask the patient to raise his or her arm to shoulder height (90° of

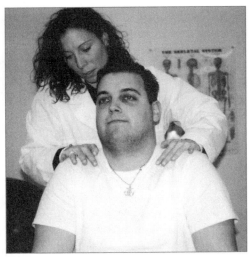

Figure 5.14 MMT: Shoulder Elevation

shoulder flexion) or available range. Tell the patient, "Don't let me push your arm down." Apply resistance to the mid-humerus toward shoulder extension. Repeat on the involved extremity (see Figure 5.15).

■ Ask the patient to raise his or her arm out to the side (90° of shoulder abduction) or available range. Tell the patient, "Don't let me push your arm down." Apply resistance to the mid-humerus into adduction. Repeat on the involved extremity (see Figure 5.16).

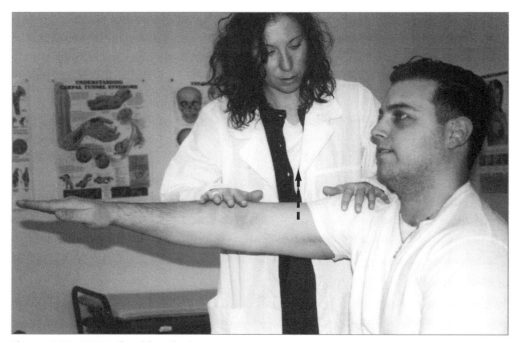

Figure 5.15 MMT: Shoulder Flexion

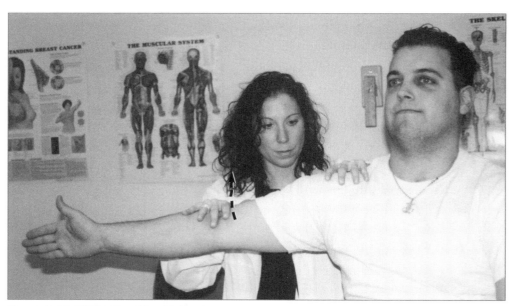

Figure 5.16 MMT: Shoulder Abduction

■ Ask the patient to maintain arm in a field goal position (90° of abduction and elbow flexion) or in available range (neutral position for screening shoulder external rotation). Place hand on the inferior aspect of the patient's triceps and the dorsal wrist surface. Tell the patient, "Don't let me push your arm down." Apply resistance toward internal rotation. Repeat on the involved extremity (see Figure 5.17).

Figure 5.17 MMT: Shoulder External Rotation

■ Ask the patient to maintain arm in a field goal position (90° of abduction and elbow flexion) or in available range (neutral position for screening shoulder internal rotation). Place hand on the inferior aspect of the patient's triceps and the dorsal wrist surface. Tell the patient, "Don't let me push your arm up." Apply resistance toward external rotation. Repeat on the involved extremity (see Figure 5.18).

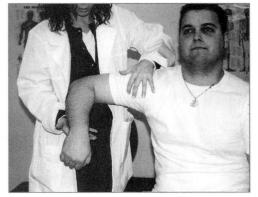

Figure 5.18 MMT: Shoulder Internal Rotation

■ Ask the patient to flex his or her elbow or touch the shoulder. Place hand on the patient's mid-forearm and the other hand supporting the arm. Tell the patient, "Don't let me pull your arm down." Apply resistance toward elbow extension. Repeat on the involved extremity (see Figure 5.19).

■ Ask the patient to flex his or her elbow or touch the shoulder. Place hand on the patient's mid-forearm and the other hand supporting the arm. Tell the patient, "Don't let me push your arm toward your chest." Apply resistance toward elbow flexion. Repeat on the involved extremity (positioning is the same as for Figure 5.15).

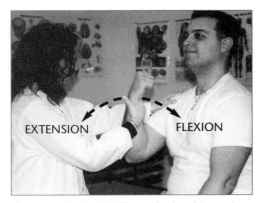

Figure 5.19 MMT: Elbow Flexion/Extension

■ Ask the patient to hold his or her arm close to body with the elbow flexed at 90°. Forearm should be placed in neutral. Position hand as though you are shaking the patient's hand. Tell the patient, "Don't let me turn your palm down toward the floor." Apply resistance toward pronation. Repeat on the involved extremity (see Figure 5.20).

■ Ask the patient to hold his or her arm close to body with the elbow flexed at 90°. Forearm should be placed in neutral. Position hand as though you are shaking the patient's hand. Tell the patient, "Don't let me turn your palm up toward the ceiling." Apply resistance toward supination. Repeat on the involved extremity (positioning is the same as for Figure 5.16).

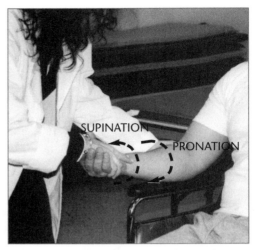

Figure 5.20 MMT: Supination/Pronation

■ Ask the patient to move his or her wrist into a neutral position, with wrist extended and palm turned to the floor. Support the extremity on a surface. Tell the patient, "Don't let me push your hand upward." Apply resistance on the palmar surface of the hand into extension. Repeat on the involved extremity (see Figure 5.21).

■ Ask the patient to move his or her wrist into a neutral position, with wrist extended and palm turned to the floor. Support the extremity on a surface. Tell the patient, "Don't let me push your hand downward." Apply resistance at the dorsal surface of the hand into flexion. Repeat on the involved extremity (positioning is the same as for Figure 5.21).

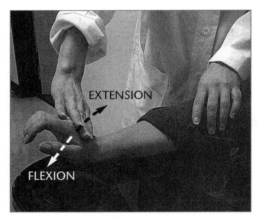

Figure 5.21 MMT: Wrist Flexion/Extension

■ Ask the patient to raise his or her knee up toward the ceiling. Tell the patient, "Don't let me push your leg down." Apply resistance to the anterior thigh toward the floor (toward hip extension). Repeat on the involved extremity (see Figure 5.22).

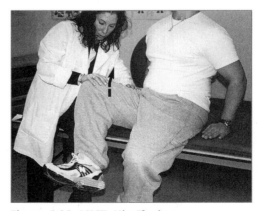

Figure 5.22 MMT: Hip Flexion

■ Ask the patient to extend his or her knee or kick out to touch your hand. Support the weight of the lower extremity. Tell the patient, "Don't let me push your lower leg down." Apply resistance to the mid-leg toward flexion. Repeat on the involved extremity (see Figure 5.23).

- Ask the patient to flex his or her knee. Support the weight of the lower extremity. Tell the patient, "Don't let me push your lower leg upward." Apply resistance to the mid-calf toward extension. Repeat on the involved extremity.

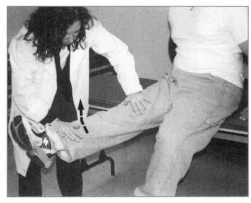

Figure 5.23 MMT: Knee Extension

- Ask the patient to point his or her toe downward. Place hand on the bottom of the patient's foot. Tell the patient, "Don't let me push your foot upward." Apply resistance toward dorsiflexion. Repeat on the involved extremity (see Figure 5.24).

- Ask the patient to point his or her toes toward the nose. Place hand on the dorsum of the patient's foot. Tell the patient, "Don't let me push your foot downward." Apply resistance toward plantar flexion. Repeat on the involved extremity (see Figure 5.25).

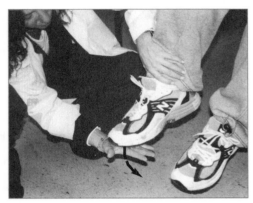

Figure 5.24 MMT: Plantar Flexion

Functional screening for FMS:

- Can the patient hold his or her head upright against gravity?

- Does the patient have the upper-extremity strength to push him- or herself up from supine to sit?

- Does the patient have the lower-extremity strength needed to perform sit to stand?

- Can the patient reach into an overhead cabinet for a 2-pound jar and open it?

- Can the patient bring a glass to his or her mouth?

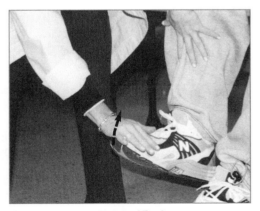

Figure 5.25 MMT: Dorsiflexion

☞ A functional muscle strength impairment is indicated if the muscle grade is not equal to that of the noninvolved extremity, is less than the represented norm grade (see Tables 5.4 and 5.5), causes an inability to move the extremity against gravity, significantly fatigues within an ADL task, or impedes the ability to use the extremity for ADL tasks.

2. *Grip strength* is the ability to maintain a hand grasp over a period of time (see Tables 5.6 and 5.7). *Note:* For clinical screening, use of a dynamometer is required.

■ Place the patient in a seated position with shoulder adducted and neutrally rotated, elbow flexed at 90°, and forearm and wrist in a neutral position.

Table 5.4 Functional Muscle Strength

Muscle Group	Activity	Functional Muscle Grade
Cervical rotators	Looking over right and left shoulders when crossing the street	Fair
Cervical flexors	Looking down when reaching into a high closet or cabinet	Trace
Cervical extensors	Looking or reaching into a high closet or cabinet	Fair
Shoulder flexors	Placing a hanging garment into a closet	Fair
Shoulder extensors	Reaching back to put arm in a sleeve	Fair+
Shoulder abductors	Laterally reaching for an item in the kitchen	Fair+
Shoulder adductors	Washing the opposite side of body	Fair
Shoulder external rotators	Brushing or shampooing hair	Poor
Shoulder internal rotators	Tucking the back of shirt into pants	Fair+
Elbow flexors	Drinking from a glass; self-feeding	Fair+
Elbow extensors	Pulling up knee-high socks	Poor+
Supinators	Brushing teeth	Fair
Pronators	Locking the door with a key	Fair
Radial and ulnar deviation	Cleaning the countertop	Fair−
Finger flexors	Maintaining grip on a bar of soap; squeezing a tube of toothpaste; writing	Fair+
Finger extensors	Releasing a beverage; releasing the arms of a chair when standing up	Fair−
Hip flexors	Negotiating steps	Fair+
Hip extensors	Rising from a seated position	Fair+
Hip abductors	Side-stepping into a bathtub	Fair
Hip adductors	Crossing lower extremities to don lower body clothing	Fair−
Knee flexors	Picking up an item off of the floor	Fair
Knee extensors	Getting out of a car	Good
Plantar flexors	Standing on toes to retrieve an item from a high shelf	Good
Dorsiflexors	Clearance of foot during the step-through phase of gait	Fair

Table 5.5 Activities Requiring Functional Range of Motion

Functional Movement	Activity	Functional ROM in Degrees
Cervical rotation	Looking over right and left shoulders for traffic when crossing	30
Cervical flexion	Looking down to button a shirt	10
Cervical extension	Looking up at a traffic light or an item on a top shelf	115
Shoulder flexion	Shampooing or combing hair	115
Shoulder extension	Reaching for the support surface during stand to sit	20
Shoulder abduction	Placing a barrette in hair	80
Shoulder adduction	Washing opposite side of body	25
Shoulder external rotation	Combing back of head; reaching behind body for shirtsleeve while dressing	80
Shoulder internal rotation	Fastening a bra	80
Elbow flexion	Washing face; bringing hand to mouth during feeding	150
Elbow extension	Reaching to don or doff lower body clothing	−20
Forearm supination	Brushing hair and teeth	15
Forearm pronation	Turning a key in the lock	15
Wrist flexion	Pulling up on a car door handle to open it	30
Wrist extension	Pushing up from a support surface	25
Ulnar/radial deviation	Opening a door; wiping down a countertop	30
Finger flexion	Buttoning; zipping; wringing out a sponge; writing; holding a razor	MP = 20, PIP = 70, DIP = 10
Finger extension	Releasing any object; pushing a door open with hand	−60
Hip flexion	Sitting on the commode; donning socks	80
Hip extension	Getting up from a chair	80
Hip abduction	Sidestepping over the bathtub	30
Hip adduction	Sitting with legs crossed	30
Knee flexion	Sitting; crossing legs to don or doff lower body clothing	80
Knee extension	Standing up from a seated position; functional ambulation	−15
Ankle dorsiflexion	Functional transfers; the step-through phase of the gait cycle	5
Ankle plantar flexion	Sliding foot into an enclosed shoe	20

Note. ROM = range of motion; MCP = metacarpalphalangeal; PIP = proximal interphalangeal; DIP = distal interphlangeal.

Table 5.6 Functional Pinch and Grip Strength

Functional Pinch and Grip	Activity	Pounds Required
Lateral pinch	Open a food package	3
Palmar/three-point pinch	Button a shirt; hold a pen	2
Tip/two-point pinch	Finger feed; plug in an appliance	3
Gross grasp	Open a jar; hold a laundry basket; drain water from a pot	10

Table 5.7 Grasp Dynamometer Norms in Pounds

Gender	Norms at Age						
	20	30	40	50	60	70	75
Male							
Right	121	122	117	113	90	75	66
Left	104	110	113	102	77	65	55
Female							
Right	70	79	70	66	55	49	42
Left	61	68	62	57	46	41	37

Note. From "Grip and Pinch Strength: Normative Data for Adults," by V. Mathiowetz, N. Kashman, G. Volland, K. Weber, M. Dowe, and S. Rogers, 1985, *Archives of Physical Medicine Rehabilitation, 66,* pp. 69–74. Copyright 1985 by Saunders. Reprinted with permission.

■ Position the dynamometer in the patient's hand.

■ Instruct the patient to squeeze as hard as he or she can (see Figure 5.26).

■ Repeat for three trials.

■ Record the average of three trials.

3. *Pinch strength* is the ability to maintain a fingertip grasp over a period of time.

Tip pinch (see Table 5.8):

Figure 5.26 Dynamometer

■ Refer to the above positioning for grip strength.

■ Position the pinch meter between the tips of the patient's thumb and index finger.

Table 5.8 Two-Point Tip Pinch Norms in Pounds

Gender	Norms at Age						
	20	30	40	50	60	70	75
Male							
Right	18	17	18	18	16	14	14
Left	17	17	18	18	15	13	14
Female							
Right	11	12	11	12	10	10	9
Left	10	12	11	11	10	10	9

Note. From "Grip and Pinch Strength: Normative Data for Adults," by V. Mathiowetz, N. Kashman, G. Volland, K. Weber, M. Dowe, and S. Rogers, 1985, *Archives of Physical Medicine Rehabilitation, 66,* pp. 69–74. Copyright 1985 by Saunders. Reprinted with permission.

■ Instruct the patient to squeeze as hard as he or she can (see Figure 5.27).

■ Repeat for three trials.

■ Record the average of three trials.

Lateral pinch (see Table 5.9):

■ Refer to the above positioning for grip strength.

■ Position the pinch meter between the pad of the patient's thumb and the lateral surface of the index finger.

■ Instruct the patient to squeeze as hard as he or she can (see Figure 5.28).

■ Repeat for three trials.

■ Record the average of three trials.

Figure 5.27 Tip Pinch

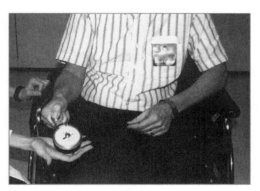

Figure 5.28 Lateral Pinch

Palmar pinch (see Table 5.10):

■ Refer to the above positioning for grip strength.

■ Position the pinch meter between the pad of the patient's thumb and the pads of the index and middle fingers.

Table 5.9 Lateral Pinch Norms in Pounds

	Norms at Age						
Gender	**20**	**30**	**40**	**50**	**60**	**70**	**75**
Male							
Right	26	26	25	27	23	19	20
Left	26	26	25	26	22	19	19
Female							
Right	17	19	17	17	15	14	12
Left	16	18	16	16	14	14	11

Note. From "Grip and Pinch Strength: Normative Data for Adults," by V. Mathiowetz, N. Kashman, G. Volland, K. Weber, M. Dowe, and S. Rogers, 1985, *Archives of Physical Medicine Rehabilitation, 66*, pp. 69–74. Copyright 1985 by Saunders. Reprinted with permission.

Table 5.10 Three-Point Palmar Pinch Norms in Pounds

	Norms at Age						
Gender	20	30	40	50	60	70	75
Male							
Right	26	25	24	24	22	18	19
Left	26	25	25	24	21	19	18
Female							
Right	17	19	17	17	15	14	12
Left	16	18	16	16	14	14	11

Note. From "Grip and Pinch Strength: Normative Data for Adults," by V. Mathiowetz, N. Kashman, G. Volland, K. Weber, M. Dowe, and S. Rogers, 1985, *Archives of Physical Medicine Rehabilitation, 66,* pp. 69–74. Copyright 1985 by Saunders. Reprinted with permission.

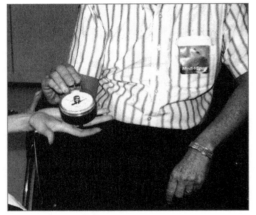

Figure 5.29 Palmar Pinch

- Instruct the patient to squeeze as hard as he or she can (see Figure 5.29).

- Repeat for three trials.

- Record the average of three trials.

SCREENING FOR FUNCTIONAL GRIP STRENGTH

- Can the patient open a large jar?

- Can the patient hold a gallon of milk?

- Can the patient open plastic containers?

SCREENING FOR BRAIN STEM CRANIAL NERVE

Refer to Section 7: "Cranial Nerve Function Screening."

SCREENING FOR CEREBELLAR AND BASAL GANGLIA MOVEMENT DISORDERS

Refer to Section 6: "Cerebellar and Basal Ganglia Function Screening" for the evaluation of the following:

- Intention tremors

- Ataxia

- Finger to nose

- Finger to finger

- Finger to therapist's finger

- Alternate patient's nose to therapist finger

- Dyssynergia

- Dysmetria

- Dysdiadochokinesia

- Rebound phenomenon

- Asthenia

- Dystonia

- Nystagmus

- Dysarthria

- Tremors at rest

- Akinesia

- Bradykinesia

- Cunctation-festinating gait

- Chorea

- Athetosis

- Hemiballismus

- Tics.

SCREENING FOR DEEP TENDON REFLEXES

Note: When testing deep tendon reflexes, hyperreflexive reactions indicate upper motor neuron injuries such as Guillain-Barré, multiple sclerosis, or other CNS deficits. Hyporeflexive reactions indicate lower motor neuron injuries such as neuropathy (e.g., radiculopathy, plexopathy, polyneuropathy, mononeuropathy) and myopathies.

1. *Biceps tendon reflex.*

- Place the patient's arm (in 90° of elbow flexion) across your forearm. The therapist's hand should be under the medial aspect of the patient's elbow.

- Place thumb on the patient's biceps tendon in the cubital fossa of the elbow. To locate the tendon, ask the patient to tense his or her biceps muscle as your thumb continues to rest in the cubital fossa. The biceps tendon will stand out under the your thumb (see Figure 5.30).

■ Instruct the patient to relax his or her arm. Maintain thumb over the patient's biceps tendon in the cubital fossa.

■ Tap your thumbnail with the narrow end of the reflex mallet. The patient's biceps should jerk slightly—a movement that the can be both felt and seen. The patient's fingers also will extend.

☞ An impairment is indicated if the biceps tendon reflex is hyperreflexive, hyporeflexive (or sluggish), or absent.

2. *Brachioradialis tendon reflex.*

■ Place the patient's arm (in 90° of elbow flexion) across your forearm. Your hand should be under the medial aspect of the patient's elbow.

■ Palpate the tendon of the brachioradialis at the distal end of the radius.

■ Tap the brachioradialis tendon with the flat edge of a reflex mallet (see Figure 5.31). The patient should exhibit a small radial jerk.

☞ An impairment is indicated if the brachioradialis and biceps tendon reflex are hyperreflexive, hyporeflexive, or absent.

3. *Triceps tendon reflex.*

■ Place the patient's arm resting in his or her lap. Place one hand on the patient's triceps muscle.

■ Palpate the tendon of the triceps muscle just above the elbow.

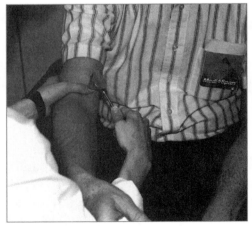

Figure 5.30 Biceps Tendon Reflex

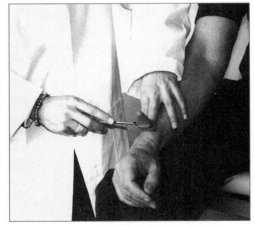

Figure 5.31 Brachioradialis Tendon Reflex

- Tap the triceps tendon with the flat edge of a reflex mallet (see Figure 5.32). The patient's elbow should extend slightly.

☞ An impairment is indicated if the triceps tendon reflex is hyperreflexive, hyporeflexive, or absent.

4. *Patellar tendon reflex.*

- Place the patient in a seated position with knees dangling (or with one leg crossed over the other).

- Palpate the soft tissue depression on either side of the tendon in order to locate it accurately and attempt to elicit the reflex by tapping the tendon with the flat edge of a reflex mallet at the level of the knee joint (see Figure 5.33). The patient's knee should move into extension.

☞ An impairment is indicated if the patellar tendon reflex is hyperreflexive, hyporeflexive, or absent.

5. *Achilles tendon reflex.*

- Ask the patient to sit with legs dangling. If the patient is lying down in supine position, flex one leg at both the hip and the knee; rotate the hip internally so that the leg rests across the opposite shin.

- Place thumb and fingers into the soft tissue depressions on either side of the patient's tendon and locate the Achilles tendon accurately (see Figure 5.34).

- Strike the tendon with the flat end of a reflex mallet to induce a sudden, involuntary plantar flexion of the foot. The patient's ankle and foot should move into plantar flexion.

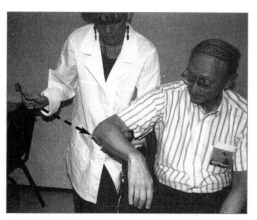

Figure 5.32 Triceps Tendon Reflex

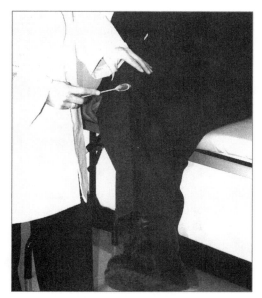

Figure 5.33 Patellar Tendon Reflex

Figure 5.34 Achilles Tendon Reflex

☞ An impairment is indicated if the Achilles tendon reflex is hyperreflexive, hyporeflexive, or absent.

6. *Plantar reflex.*

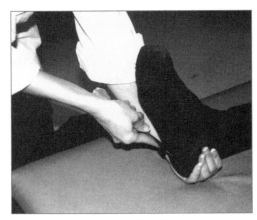

Figure 5.35 Plantar Reflex Test

■ Place the patient in supine position.

■ Use a firm object (e.g., a key or a wooden tongue depressor) to stroke the patient's foot on the plantar surface. Stroke the foot on the lateral aspect, starting from the heel and moving upward to the ball of the foot. At the ball of the foot begin curving across the foot to the medial aspect (see Figure 5.35). The patient's big toe should flex.

☞ An impairment is indicated if the plantar reflex is hyperreflexive, hyporeflexive, or absent.

SCREENING FOR PRIMITIVE REFLEXES AND REACTIONS

Primitive reactions are reflexes that humans are born with or that develop and become integrated by the CNS in infancy and toddlerhood. These reactions facilitate gross motor patterns of flexion and extension. Adults who exhibit primitive reactions usually have sustained severe brain damage.

Spinal level reflexes are basic to mobility patterns. These movement patterns are dominated by flexion or extension. Spinal reflexes are integrated within the first 2 months of life.

1. *Placing reaction of the upper limb.*

■ Position the patient in either sitting or supine.

■ Apply a brushing stimulus to the dorsum of the patient's hand.

■ Observe what occurs at the limb when donning a shirt or pulling off bed covers.

☞ Impairment is indicated if the patient exhibits flexion of the stimulated arm. Functionally, a positive reaction can impede normal movement (upper-extremity extension), as tone is increased by the stimulation of the sleeve placed over the patient's limb.

2. *Moro reflex.*

■ Place the patient in any position.

■ Apply a loud noise or sudden movement of the patient's support surface.

■ Observe the patient's postural responses in loud environments or during therapy.

☞ Impairment is indicated if the patient exhibits a startled response or abduction, extension, and external rotation of the arms. Functionally, this reflex can significantly impede bed or wheelchair positioning.

3. *Grasp reflex.*

- Place the patient in any position.

- Apply pressure in the palm of the patient's hand from the ulnar side.

- Observe the patient's postural response when the hand is used for weight bearing.

☞ Impairment is indicated if the stimulus causes a grasp response, with the fingers flexing and adducting. Functionally, this reflex will inhibit the patient's ability to grasp and release eating utensils, to grasp and release clothing during dressing, and to push up from the wheelchair.

4. *Flexor withdrawal.*

- Position the patient either in supine or sitting with legs fully extended.

- Apply a brushing movement to the sole of the patient's foot.

☞ Impairment is indicated if the stimulated leg postures into flexion (see Figure 5.36). Functionally, this reflex can impede the patient's ability to accept weight onto the lower extremities during transitional movements and inhibit his or her ability to perform standing ADL tasks or functional ambulation.

Figure 5.36 Flexor Withdrawal

5. *Crossed extension.*

- Position the patient in supine with one leg in extension and the other leg fully flexed.

- Passively flex the extended leg.

☞ Impairment is indicated if the stimulus causes the flexed leg to extend and causes hip adduction and internal rotation (see Figure 5.37). Functionally, patients may have difficulty with bed bridging, anterior weight shifts in preparation for transfers, and balance deficits impeding standing ADL tasks and ambulation.

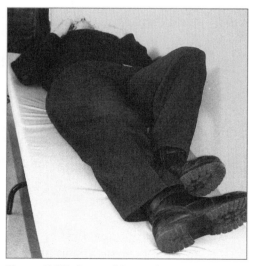

Figure 5.37 Crossed Extension

Brain stem reflexes are static reflexes directly affecting postural tone. These reflexes occur in response to changes in head position or body position in relation to the head. Brain stem reflexes are integrated within the first 6 months of life.

1. *Asymmetrical tonic neck reflex (ATNR).*

- Position the patient in supine or sitting with arms and legs extended.

- Passively turn the patient's head 90° to one side.

☛ Impairment is indicated if the patient assumes flexion of the extremities toward the skull side and extension of the extremities on the face side (see Figure 5.38). Functionally, ATNR may prevent dissociation of head and neck visual scanning, rolling in bed, or holding objects with both hands and bringing objects to the mouth. Sitting balance may be decreased as the patient's extremities are flexed unilaterally.

2. *Symmetrical tonic neck reflex (STNR).*

- Place the patient in a seated position.

- Passively flex the patient's head toward his or her chest.

- Passively extend the patient's head.

- Observe the postural response during supine to sit or during an anterior weight shift.

☛ Impairment is indicated if flexion of the upper extremities and extension of the lower extremities is present when the patient's head is flexed (see Figure 5.39) and when extension of the upper extremities and flexion of the lower extremities is noted on cervical extension (see Figure 5.40). Functionally, the patient with a positive STNR will demonstrate difficulty transitioning from supine to sitting, causing flexion of the upper extremities and extension of the

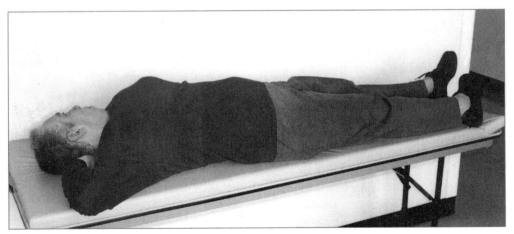

Figure 5.38 ATNR in Supine

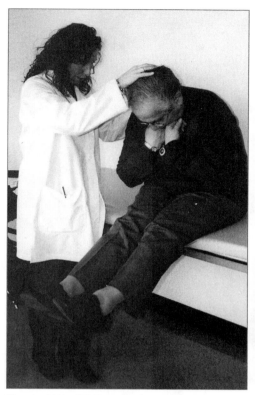

Figure 5.39 STNR in Sitting with Head Flexed

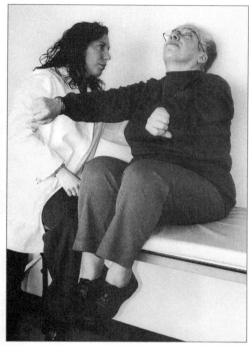

Figure 5.40 STNR in Sitting with Head Extended

lower extremities as he or she lifts the head. Risk of falls during transfers is evident, as the patient may exhibit lower-extremity buckling when anteriorly shifting weight.

3. *Tonic labyrinthine reflex (TLR).*

- Position the patient in supine.

- Observe the postural response.

- Position the patient in prone.

- Observe the postural response.

☞ Impairment is indicated if the patient exhibits extensor tone while positioned supine (see Figure 5.41) and flexor tone while positioned prone (see Figure 5.42). Functionally, the TLR prevents the patient from transitioning between movements when initiated with head movements or sitting up unassisted. For example, supine to sit may cause a patient to shoot into a full extensor pattern, making movement impossible. Patients seated in wheelchairs for increased periods may present with extensor tone that limits their ability to communicate and visually scan their environment, and that increases the risk of sliding out of the wheelchair.

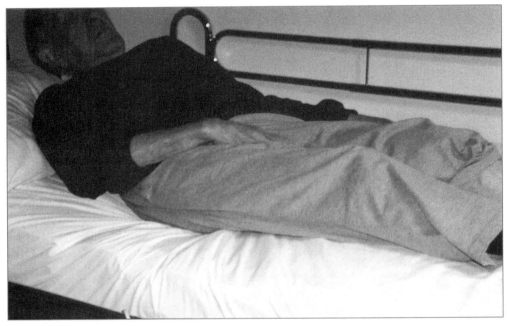

Figure 5.41 TLR in Supine

Figure 5.42 TLR in Prone

4. *Positive supporting reactions.*

- Position the patient in supine, sitting, or standing.

- Apply firm contact to the ball of the patient's foot or place the foot on the floor.

- Observe the patient's postural change during sitting and transfers.

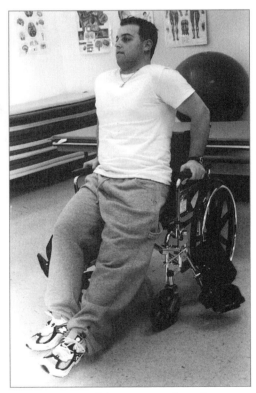

Figure 5.43 Positive Support Reaction

☞ Impairment is indicated if the patient exhibits extensor tone throughout the leg being stimulated (see Figure 5.43). Functionally, patients may have difficulty shifting weight onto the lower extremities in preparation for transfers, standing ADL tasks, functional ambulation, and step negotiation.

5. *Associated reactions.*

- Place the patient in any position.

- Have the patient squeeze any object or apply resistance to any motion.

☞ Impairment is indicated if the motion required for the stimulus is mimicked on the opposite extremity. For example, a patient given a brush to hold will mirror the same grasping pattern on the nonstimulated extremity. Functionally, patients with positive associated reactions may demonstrate increased tone during yawning, sneezing, or bathing, making limb use difficult.

Midbrain reactions facilitate the development of maturationally acquired motor milestones. These reactions allow one to automatically right the head in relation to the body and to space. Such reactions facilitate an individual's spatial orientation during movements and activities. Midbrain reactions are apparent in childhood and remain throughout the life span.

1. *Righting reactions.*

- Position the patient in supine with arms and legs fully extended.

- Passively rotate the patient's head to one side and maintain its position.

- Observe the patient during all transitional movements.

☞ Impairment is indicated if the patient rotates his body "loglike" or segmentally toward the direction of the rotated head (see Figure 5.44). Functionally, the patient may demonstrate difficulty getting out of bed and sitting up.

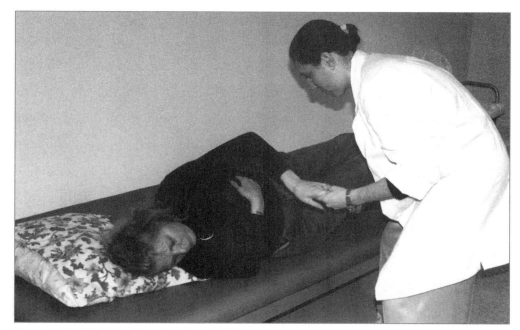

Figure 5.44 Righting Reaction: Log Roll

Basal ganglial level reflexes, or *equilibrium reactions,* enable individuals to adapt to changes in the body's center of gravity. Equilibrium reactions are achieved through the integration of vestibular, visual, and tactile input. These reactions begin at 6 months and continue throughout life. Such reactions allow for fall prevention when an individual's balance and visual orientation are challenged. *Note:* The therapist should guard the patient to protect him or her from falling.

1. *Protective extension (parachute).*

- Place the patient in functional positions, such as sitting on the edge of the bed or standing during a kitchen activity.

- Push the patient to the involved side using moderate force. (The patient should feel as though he or she is going to fall.)

☞ Impairment is indicated if the patient does not extend his or her arms and hands to break the fall.

SCREENING FOR CORTICAL LEVEL REFLEXES

1. *Optic righting.*

- Position the patient in sitting with head laterally flexed. The patient's eyes must be open.

- Present the patient with an object or a landmark in relation to space. Hold an object horizontally aligned in front of the patient. The patient should orient his or her head position in direct alignment with the object.

☞ Impairment is indicated if the patient fails to orient his or her head perpendicular with the environmental clue. Functionally, optic righting maximizes orientation in space, balance, and visual ability.

SCREENING FOR BALANCE

Balance is observed during transitional movements and functional tasks.

- Is the patient able to perform the task without loss of balance?

- What is the quality of the movement?

- What does the patient's trunk look like?

- Which positions make the patient's balance deteriorate?

- Does the patient's balance deteriorate with rapid directional changes or increased external stimuli?

- What happens to the patient's balance when he or she moves his or her center of gravity over his or her base of support?

- Does the patient's balance change with head or eye movement, lighting, or changes in support surfaces?

- If decreased balance is noted, does the patient experience vertigo or complain that the room is spinning or becoming dark and blurry?

☞ Impairment is indicated if the patient demonstrates loss of balance either in sitting or standing during functional activities.

RED FLAGS
SIGNS AND SYMPTOMS OF MOTOR DYSFUNCTION

A red flag is an indicator or symptom of dysfunction. If a patient is observed displaying any of the following red flags, it is likely that the patient possesses motor impairments.

CEREBELLAR DYSFUNCTION

- Ataxic movements impeding a patient's ability to bring a cup or spoon to the mouth without spilling
- Over- or undershooting when reaching for items in a closet
- Wide-based staggering gait during functional activities
- Noted nystagmus with functional complaints of dizziness, diplopia, and difficulty reading and writing
- Inability or difficulty applying make-up secondary to intension tremors
- Difficulty or inability to clap hands
- Hypertonia or hypotonia
- Broken or slurred speech.

BASAL GANGLIA DYSFUNCTION

- Pill rolling or tremorlike movement at rest
- Difficulty bringing an eating utensil to the mouth or both hands toward the face in grooming secondary to upper-extremity rigidity
- Inability to initiate voluntary movement
- Increased time noted to perform motor components of self-care or mobility due to bradykinesia
- Jerky movement inhibiting fine and gross motor tasks.

BRAIN STEM DYSFUNCTION

- Complaints of vertigo (i.e., the room spinning)
- Loss of balance during seated and standing ADL tasks
- Diplopia or disconjugate gaze

- Flaccidity of an extremity
- Spasticity of an extremity
- Inability to perform bed mobility secondary to extensor tonal patterns
- Difficulty performing anterior weight shifts secondary to extensor tone dominance
- Hypertonicity with yawning or coughing
- Associated reactions.

PRIMARY MOTOR AREA

- Inability to voluntarily move the involved extremity
- Spasticity.

MOTOR TRACT DYSFUNCTION

- Inability to initiate and carry out a desired movement
- Decreased visual scanning, binocular vision, and diplopia
- Difficulty swallowing
- Dysarthric speech
- Inability to coordinate head and neck movements during reaching, scanning, and ADL performance
- Difficulty ambulating on uneven terrain secondary to loss of balance
- Inability or difficulty transitioning from picking up low items to returning to normal midline trunk in sitting
- Inability to maintain head in a neutral position or hold head up against gravity.

FRONTAL LOBE DYSFUNCTION

- Inability to correctly select or use eating utensils secondary to apraxia
- Inability to perform grooming and hygiene secondary to incorrect utensil use (or apraxia).

UPPER MOTOR LESION DYSFUNCTION

- Increased tone inhibiting the ability to perform ADL tasks.

RANGE OF MOTION DYSFUNCTION

- Inability to fasten rear-closing garment
- Inability to bend to don socks and shoes
- Inability to use hand to push off of support surface.

TONAL DYSFUNCTION

- Difficulty or posturing when moving between positions

- Posturing of an extremity noted during strenuous movements

- Complaints of a "heavy" arm

- Difficulty eating secondary to difficulty with smooth upper-extremity transitions to mouth.

FUNCTIONAL MUSCLE STRENGTH DYSFUNCTION

- Difficulty transitioning from sit to stand

- Difficulty lifting a heavy container

- Inability to open jars

- Difficulty carrying items.

AVAILABLE IN-DEPTH ASSESSMENTS

If motor screening detects impairment, the following in-depth assessments are available:

Activities-Specific Balance Confidence Scale (ABC)
L. E. Powell, A. M. Myers
Adults with balance deficits
Powell, L. E., & Myers, A. M. (1995). The Activities-Specific Balance Confidence (ABC) Scale. *Journal of Gerontology, 50A*(1), M28–M34.

Arm Motor Ability Test (AMAT)
B. Kopp
Adults with neuromuscular and musculoskeletal disorders
Kopp, B., Kunkel, A., Floor, H., Platz, T., Rose, U., & Mauritz, K. (1997). The Arm Motor Ability Test: Validity and sensitivity to change of an instrument for assessing disabilities in activities of daily living. *Archives of Physical Medicine and Rehabilitation, 78,* 615–620.

Assessment of Motor and Process Skills (AMPS)
A. G. Fisher
Children and adults with developmental, neurologic, and musculoskeletal disorders
Anne G. Fischer, ScD, OTR/L, FAOTA
AMPS Project
Occupational Therapy Bldg.

Colorado State University
Fort Collins, CO 80523

Barthel Index
F. I. Mahoney, D. W. Barthel
Adults with neuromuscular or musculoskeletal disorders
Mahoney, F. I., & Barthel, D. W. (1965). The Barthel Index. *Maryland State Medical Journal, 14,* 61–65.

Berg Balance Scale (Berg)
K. O. Berg, S. L. Wood-Dauphinee, J. I. Williams, B. Maki
Adults with balance deficits
Berg, K. O., Wood-Dauphinee, S. L., Williams, J. I., & Maki, B. (1992). Measuring balance in the elderly: Validation of an instrument. *Canadian Journal of Public Health, 83,* S7–S11.

Box and Block Test of Manual Dexterity
V. Mathiowetz, G. Volland, N. Kashman, K. Weber
Adults with fine motor deficits secondary to neurologic and/or musculoskeletal impairment
Mathiowetz, V., Volland, G., Kashman, N., & Weber, K. (1985). Adult norms for the Box and Block Test of Manual Dexterity. *American Journal of Occupational Therapy, 39,* 386–391.

Canadian Neurological Scale (CNS)
R. Cote, R. N. Battista, C. Wolfson
Adults in the acute stage of cerebrovascular accident
Cote, R., Battista, R. N., & Wolfson, C. (1989). The Canadian Neurological Scale: Validation and reliability assessment. *Neurology, 39,* 638–643.

Canadian Occupational Performance Measure (COPM)
M. Law, S. Baptiste, A. Carswell, M. A. McColl, H. Polatajko, N. Pollock
Children and adults with cognitive, psychosocial, neurologic, and/or musculoskeletal impairment
Canadian Association of Occupational Therapists
110 Eglington Ave. West, 3rd Floor
Toronto, ON M4R 1A3 Canada

Clinical Test of Sensory Interaction in Balance (CTSIB)
M. Watson
Adults with balance deficits
Watson, M. (1992). Clinical Test of Sensory Interaction in Balance. *Physiotherapy Theory and Practice, 8,* 176–178.

Dizziness Handicap Inventory (DHI)
G. P. Jacobson, C. W. Newman
Adults with balance deficits secondary to vestibular impairment
Jacobson, G. P., & Newman, C. W. (1990). The development of the Dizziness Handicap Inventory. *Archives of Otolaryngology, Head, and Neck Surgery, 116,* 424–427.

Emory Test of Motor Function
S. Wolf
Adults with neurologic insult
Wolf, S. L., Lecraw, D. E., Barton, L. A., & Jann, B. B. (1989). Forced use of the hemiplegic upper extremity to reverse the effects of learned nonuse among chronic stroke and head-injured patients. *Experimental Neurology, 104,* 125–132.

Falls Efficacy Scale (FES)
M. E. Tinnetti, D. Richman, L. Powell
Elderly persons with balance deficits
Tinnetti, M. E., Richman, D., & Powell, L. (1990). Falls efficacy as a measure of fear of falling. *Journal of Gerontology, 45,* P239–P243.

Fugl-Meyer Assessment of Sensorimotor Recovery After Stroke
A. R. Fugl-Meyer
Adults who sustained cerebrovascular accident
Fugl-Meyer, A. R. (1980). Post-stroke hemiplegia assessment of physical properties. *Scandinavian Journal of Rehabilitation Medicine, 7,* 85–93.

Functional Independence Measure (FIM)
Center for Functional Assessment Research at State University of New York at Buffalo
Director: Carl V. Granger, MD
Adults with neurologic and/or musculoskeletal impairment
Uniform Data System for Medical Rehabilitation
232 Parker Hall
State University of New York at Buffalo
3435 Main St.
Buffalo, NY 14214

Functional Reach Test (FRT)
P. W. Duncan, D. K. Weiner, J. Chandler, S. A. Studenski
Adults with balance deficits secondary to neurologic damage
Duncan, P. W., Weiner, D. K., Chandler, J., & Studenski, S. A. (1990). Functional reach: A new clinical measure of balance. *Journal of Gerontology, 45,* 192–197.

Functional Test for the Hemiparetic Upper Extremity
D. J. Wilson, L. L. Baker, J. A. Craddock
Adults with hemiplegia
Los Amigos Research and Education Institute, Inc.
PO Box 3500
Downey, CA 90242

Jebsen Hand Function Test
R. H. Jebsen, N. Taylor, R. B. Trieschmann, M. J. Trotter, L. A. Howard
Children and adults with neurologic and/or musculoskeletal impairment
Sammons Preston Inc.
PO Box 50710
Bolingbrook, IL 60440-5071
(800) 323-5547

Katz Index of ADL
S. Katz, A. B. Ford, R. W. Moskowitz,
B. A. Jackson, M. W. Jaffe
Elderly persons with neurologic and/or musculoskeletal impairment
Katz, S., Ford, A. B., Moskowitz, R. W., Jackson, B. A., & Jaffe, M. W. (1963). The Index of ADL: A standardized measure of biological and psychological function. *Journal of the American Medical Association, 185,* 914–919.

Klein–Bell Activities of Daily Living Scale
R. M. Klein, B. J. Bell
Infants to adulthood—neurologic and/or musculoskeletal impairment
Health Sciences Center for Educational Resources
University of Washington
T-281 Health Science Bldg.
Box 357161
Seattle, WA 98195-7161

Kohlman Evaluation of Living Skills (KELS)
L. Kohlman-Thomas
Adolescents and adults with psychiatric, developmental, and/or neurologic impairment
American Occupational Therapy Association
4720 Montgomery Ln.
PO Box 31220
Bethesda, MD 20824-1220

Minnesota Rate of Manipulation Tests (MRMT) and Minnesota Manual Dexterity Test (MMDT)
University of Minnesota Employment Stabilization Research Institute
Adults with neurologic and/or musculoskeletal impairment
American Guidance Services, Inc.
Publishers' Bldg.
Circle Pines, MN 55014

Morse Fall Scale (MFS)
J. M. Morse
Adults with balance deficits
Janice M. Morse
Pennsylvania State University
School of Nursing
201 Health and Human Development East
University Park, PA 16802-6508

Motor Assessment Scale (MAS)
J. H. Carr, R. B. Shepard
Adults who sustained cerebrovascular accident
Carr, J. H., & Shepard, R. B. (1985). Investigation of a new motor assessment scale for stroke patients. *Physical Therapy, 65,* 195–197.

Nine-Hole Peg Test
V. Mathiowetz, K. Weber, N. Kashman, G. Volland
Adults with fine motor coordination deficits secondary to neurologic and/or musculoskeletal impairment
Mathiowetz, V., Weber, K., Kashman, N., & Volland, G. (1985). Adult norms for the Nine Hole Peg Test of finger dexterity. *Occupational Therapy Journal of Research, 5,* 24–38.

O'Connor Finger and Tweezer Dexterity Tests
J. O'Connor
Adults with fine motor coordination deficits secondary to neurologic and/or musculoskeletal impairment
Sammons Preston Inc.
PO Box 5071
Bolingbrook, IL 60440-5071
(800) 323-5547

Physical Performance Test (PPT)
D. B. Reuben, A. L. Siu
Elderly persons with balance deficits
Reuben, D. B., & Siu, A. L. (1990). An objective measure of physical function of elderly outpatients: The Physical Performance Test. *Journal of the American Geriatric Society, 38,* 1105–1112.

Purdue Pegboard
J. Tiffin
Adults with fine motor and eye–hand coordination deficits secondary to neurologic and/or musculoskeletal impairment
Lafayette Instrument Company
3700 Sagamore Pkwy. North
PO Box 5729
Lafayette, IN 47903

Roeder Manipulative Aptitude Test
W. S. Roeder
Adolescents and adults with fine motor and eye–hand coordination deficits secondary to neurolgic and/or musculoskeletal impairment
Lafayette Instrument Company
3700 Sagamore Pkwy. North
PO Box 5729
Lafayette, IN 47903

Structured Observational Test of Function (SOTOF)
A. Laver, G. E. Powell
Elderly persons with dementia or other neurologic impairment
NFER-NELSON Publishing Company
Darville House
2 Oxford Rd. East
Windsor, Berkshire SL4 1DF England

TD (Tardive Dyskinesia) Monitor (TD Monitor)
W. Glazer
Patients receiving chronic neuroleptic maintenance
Multi-Health Systems, Inc.
908 Niagara Falls Blvd.
North Tonawanda, NY 14120-2060

Timed Get Up and Go (TGUG)
D. Podsiadlo, S. Richardson
Adults with balance deficits
Podsiadlo D., & Richardson, S. (1991). The Timed Get Up and Go: A test of basic functional mobility for frail elderly persons. *Journal of the American Geriatric Society, 39,* 142–148.

Tinnetti Balance Test of the Performance-Oriented Assessment of Mobility Problems (Tinnetti)
M. E. Tinnetti
Adults with balance deficits
Tinnetti, M. E. (1986). Performance-oriented assessment of mobility in elderly patients. *Journal of the American Geriatric Society, 34,* 119–126.

Trunk Control Test
F. P. Franchignoni, L. Tesio, C. Ricupero, M. T. Martino
Adults who sustained cerebrovascular accident
Franchignoni, F. P., Tesio, L., Ricupero, C., & Martino, M. T. (1997). Trunk Control Test as an early predictor of stroke rehabilitation outcome. *Stroke, 28,* 1382–1385.

For further information about the these assessments, refer to the following resources:

Asher, I. E. (1996). *Occupational therapy assessment tools: An annotated index* (2nd ed.). Bethesda, MD: American Occupational Therapy Association.
Impara, J. C., Plake, B. S., & Murphy L. L. (Eds.). (1998). *The thirteenth mental measurements yearbook.* Lincoln: University of Nebraska Press.
Murphy, L. L., Impara, J. C., & Plake, B. S. (Eds.). (1999). *Tests in print: An index to tests, test reviews, and the literature on specific tests* (Vols. 1 & 2). Lincoln: University of Nebraska Press.
Plake, B. S., Impara, J. C., & Murphy, L. L. (Eds.). (1999). *The supplement to* The Thirteenth Mental Measurements Yearbook. Lincoln: University of Nebraska Press.

References
Bohannon, R. W., & Smith, R. B. (1987). Interrater reliability of a modified Ashworth Scale of Muscle Spasticity. *Physical Therapy, 67,* 207.
Boone, D. C., & Azen, S. P. (1979). Normal range of motion of joints in male subjects. *Journal of Bone and Joint Surgery, 61,* A756.
Brunnstrom, S. (1941). Muscle group testing. *Physiotherapy Review, 21,* 3–21.
Krusen, F. H., & Kottke, F. J. (Eds.). (1971). Goniometry: The measurement of joint motion. *Handbook of Physical Medicine and Rehabilitation, 2nd ed.*
Lilienfeld, A. M., Jacobs, M., & Willis, M. A. (1954). A study of the reproducibility of muscle testing and certain other aspects of muscle scoring. *Physical Therapy Review, 34,* 279–289.
Mathiowetz, V., Kashman, N., Volland, G., Weber, K., Dowe, M., & Rogers, S. (1985). Grip and pinch strength: Normative data for adults. *Archives of Physical Medicine Rehabilitation, 66,* 69–74.

FUNCTIONAL MOTOR SCREENING

Patient Name _____   Date _____

Diagnosis: _____

Observation	Yes	No
Are there changes in posture/tone noted?	☐	☐
Do movements appear effortful or are tonal changes noted?	☐	☐
Does the patient have head control?	☐	☐
Can the patient sit unsupported?	☐	☐
Can the patient dissociate head and trunk?	☐	☐
Can the patient control trunk with UE use?	☐	☐
Can the patient isolate upper from lower trunk?	☐	☐
Are tremors noted during movement?	☐	☐
Can the patient perform weight shifts? Tone noted?	☐	☐
Are there tonal patterns noted during ambulation?	☐	☐
Are there extraneous movements noted during functional ambulation?	☐	☐

Range of Motion	Passive	Active	Comment
Head rotation _____	☐	☐	_____
Cervical lateral flexion	☐	☐	_____
Shoulder elevation/depression	☐	☐	_____
Shoulder retraction/protraction	☐	☐	_____
Shoulder flexion/extension	☐	☐	_____
Shoulder abduction/adduction	☐	☐	_____
Shoulder INT/EXT rotation	☐	☐	_____
Elbow flexion/extension	☐	☐	_____
Forearm supination/pronation	☐	☐	_____
Wrist flexion/extension	☐	☐	_____
Wrist ulnar/radial deviation	☐	☐	_____
Hand	☐	☐	_____
Hip flexion/extension	☐	☐	_____
Knee flexion/extension	☐	☐	_____
Ankle	☐	☐	_____

Functional observation: _____
_____

Tone	Normal	Impaired	Comment
Cervical region	☐	☐	_____
Shoulder	☐	☐	_____
Elbow	☐	☐	_____
Wrist	☐	☐	_____
Hand	☐	☐	_____
Hip	☐	☐	_____
Knee	☐	☐	_____
Foot	☐	☐	_____

Functional observation: _____

Continued

Functional Muscle Strength	WFL	Impaired	Comment
Scapula	☐	☐	_____
Shoulder	☐	☐	_____
Elbow	☐	☐	_____
Wrist	☐	☐	_____
Hand	☐	☐	_____
Hip	☐	☐	_____
Knee	☐	☐	_____
Foot	☐	☐	_____

Functional observation: _____

Pinch and Grip Strength	WFL	Impaired	Comment
Grip strength	☐	☐	_____
Tip pinch	☐	☐	_____
Lateral pinch	☐	☐	_____
Palmar pinch	☐	☐	_____

Functional observation: _____

Deep Tendon Reflexes	Hyperreflexive	Absent
Biceps	☐	☐
Brachioradialis	☐	☐
Triceps	☐	☐
Patellar	☐	☐
Achilles	☐	☐
Plantar	☐	☐

Reflexes and Reactions	Positive	Negative
Placing of the upper extremity	☐	☐
Moro reflex	☐	☐
Grasp reflex	☐	☐
Flexor withdrawal	☐	☐
Crossed extension	☐	☐
ATNR	☐	☐
STNR	☐	☐
Tonic neck labyrinthine reflex	☐	☐
Positive supporting reaction	☐	☐
Associated reactions	☐	☐
Righting reactions	☐	☐
Equilibrium reactions	☐	☐
Protective extension	☐	☐
Optic righting	☐	☐

Functional observation: _____

Continued

Balance	Intact	Impaired	Comment
Supported Sitting			
Static	☐	☐	_____
Dynamic	☐	☐	_____
Unsupported Sitting			
Static	☐	☐	_____
Dynamic	☐	☐	_____
Standing			
Static	☐	☐	_____
Dynamic	☐	☐	_____

Cerebellar and Basal Ganglia Function
SCREENING
(Postural Control, Balance, and Automated Movements)

FUNCTIONAL IMPLICATIONS OF CEREBELLAR AND BASAL GANGLIA DISORDERS

The *cerebellum* plays a role in the coordination of movement, the maintenance of posture, and equilibrium. Its major function is *proprioception*—the awareness of one's body in space. The cerebellum could be called an *error-correcting device for the motor system*. It receives proprioceptive information from the muscles and joints and sends back information to modify muscle and joint activity for the achievement of *precision motor control*.

For patients with cerebellar lesions, activities of daily living (ADL) that require fine-precision, end-point movement become severely disrupted. For example, bringing a fork to one's mouth for eating becomes a skill disrupted by *intention tremors* and *ataxia*. Similarly, drinking without spilling liquid down one's face, brushing one's hair, shaving, and placing a key in a lock to open a door all become tasks of great difficulty and frustration. Because cerebellar lesions disrupt the timing, rate, and force of movements, patients often over- or undershoot their reach for objects; therefore, attempting to reach for and grasp a drinking glass becomes a task requiring great concentration. The wide-base, unsteady, staggering ataxic gait characteristic of cerebellar lesions makes unassisted walking difficult and places patients at an increased safety risk for falls. Crossing a heavily trafficked street becomes a dangerous activity, as does walking across unlevel terrain and negotiating steps and curb cuts. *Posterior cerebellar lobe lesions* that cause *dysarthria* or a staccato voice (broken speech) make communication laborious and frustrating and cause patients to feel self-conscious when interacting with others.

The *basal ganglia* play a role in stereotypic and automated movement patterns, that is, movement patterns generated by the human nervous system at a developmentally appropriate, age-specific period (e.g., reciprocal arm swing, crawling, walking) and movement patterns learned first at a cortical level and then subcortically integrated by the basal ganglia structures (e.g., riding a bike, writing). Patients with basal ganglial lesions have difficulty initiating and terminating movements; can experience *rigidity* or *dystonia* that severely limit movement; and commonly implement involuntary, undesired movements, such as tics, tremor, and chorea. The *cunctation-festinating gait* characteristic of *Parkinson's disease* (a basal ganglial disorder) disrupts the patient's ability to start and stop walking at will. In fact, the ability to start and stop movements required in all ADL is disrupted. Dressing becomes a task that may require several hours as a result of the patient's struggle to initiate voluntary movement. Rigidity and hypertonicity further limit voluntary movement in all ADL, making the most basic self-care tasks difficult and onerous. The involuntary extraneous movements of *hemiballismus, athetosis, chorea,* and *tics* (all basal ganglial disorders) are fatiguing and cause patients to feel self-conscious in public settings.

CEREBELLAR AND BASAL GANGLIA SCREENING PROCEDURES

Patients with cerebellar and/or basal ganglia disorders are screened for coordination in daily activities. Coordination screens are divided into two primary categories: *gross motor activities* and *fine motor activities*. Gross motor activities assess the patient's posture, balance, and extremity movements involving large muscle groups, for example,

- Crawling, kneeling, standing, walking, and running

- Bending down to reach for a food item in a low cabinet

- Reaching overhead for an item in a high cabinet

- Carrying a meal tray while walking

- Washing one's hair while standing in the shower.

Fine motor activities assess extremity movements involving small muscle groups, for example,

- Shaving

- Brushing one's teeth

- Slicing fruit

- Bringing a spoon to one's mouth

- Using a key to open a door.

Coordination screens are further divided into *nonequilibrium* and *equilibrium* screens. Nonequilibrium screens assess both static and dynamic movements when the body *is not* in an upright standing position. Equilibrium screens assess both static and dynamic movement when the body *is* in an upright standing position. When screening a patient with balance and coordination problems, the therapist should carefully guard *(contact guard)* the patient to reduce the risk of falls and injury. While observing the patient's performance during the screening procedures, the therapist should consider the following questions:

- Are the patient's movements smooth and precise, or broken and inaccurate?

- Is the patient able to implement voluntary movement within a reasonable or normal amount of time, or does it take the patient a considerable amount of time to implement the desired movements?

- Can the patient maintain a specific body position/posture without swaying, leaning toward one side, experiencing tremors, or making extraneous movements?

- When the patient's vision is occluded, does the quality of movement change? In other words, is the patient using visual cues to compensate for motor deficits?

CEREBELLAR SCREENING PROCEDURES

1. *Intention tremors* occur during voluntary movement of a limb and tend to increase as the limb nears its intended goal. For example, the patient may experience increased tremors as he or she attempts to use the hand to bring a spoon to the mouth. *Tremor* is the rhythmic oscillation of joints caused by alternating contractions of opposing muscle groups. Intention tremors tend to diminish or stop when the patient's limbs are at rest.

- Place the patient in a seated position.

- Give the patient a spoon and a bowl of water.

- Ask the patient to bring a spoonful of water to his or her mouth (a small portion of gelatin, pudding, or applesauce can be substituted for the water).

☛ Impairment is indicated if the patient's upper extremity tremors as he or she brings the spoon to the mouth.

Finger to nose:

- Place the patient in a seated position with upper extremities abducted to 90° (with elbows extended).

- Instruct the patient to touch his or her nose alternately with the left and right index fingers. The patient performs the task first with eyes open, then with eyes closed (see Figure 6.1).

☛ Impairment is indicated if the patient exhibits increased intention tremors as he or she attempts to touch his or her nose with each index finger. Impairment also is indicated if the patient's accuracy of movement becomes worse with eyes closed.

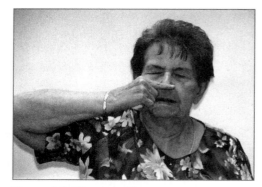

Figure 6.1 Finger to Nose

Finger to finger:

- Place the patient in a seated position.

- Instruct the patient to abduct both upper extremities to 90° (with elbows extended).

- Ask the patient to bring both hands toward the midline and touch the index fingers together. The patient performs the task first with eyes open, then with eyes closed (see Figure 6.2).

Figure 6.2 Finger to Finger

☛ Impairment is indicated if the patient experiences increased intention tremors as he or she attempts to touch the index fingers together. Impairment also is indicated if the patient's accuracy of movement becomes worse with eyes closed.

Finger to therapist's finger:

- Sit opposite the patient, who also is seated.

- Hold index finger in front of the patient and instruct the patient to touch the tip of your finger with his or her own index finger.

- Change the position of your finger during screening to assess the patient's ability to change the distance, direction, and force of movement (see Figure 6.3).

☛ Impairment is indicated if the patient has difficulty touching the therapist's finger as a result of increased intention tremors.

Alternate patient's nose to therapist's finger:

- Sit opposite the patient, who also is seated.

- Hold index finger in front of the patient and instruct the patient to alternately touch the tip of your finger and the tip of his or her own nose.

- Change the position of your finger during screening to assess the patient's ability to change the distance, direction, and force of movement (see Figure 6.4).

☛ Impairment is indicated if the patient exhibits increased intention tremors as he or she attempts to alternatively touch his or her nose and the therapist's fingertip.

Figure 6.3 Finger to Therapist's Finger

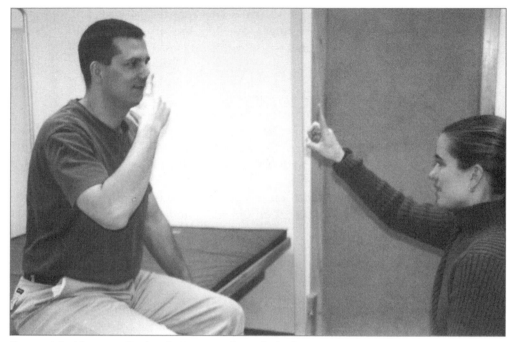

Figure 6.4 Alternate Patient's Nose to Therapist's Finger

2. *Movement decomposition* or *dyssynergia* can be assessed using the above screening procedures. Movement decomposition is characterized by movements that are broken up into their component parts rather than occurring as a smooth, single movement. For example, when the patient is instructed to touch his or her fingertip to his or her nose, he or she may first flex the elbow, then flex the shoulder, and finally adjust the position of the wrist and fingers.

☞ Impairment is indicated by movement that occurs in steps at each involved joint rather than as a singular coordinated movement occurring simultaneously at each involved joint.

3. *Dysmetria* is an inability to judge the distance and range of a movement. It is characterized by overshooting *(past pointing)* or undershooting one's reach for a target object.

Have the patient complete one of the following two screening procedures.

Procedure 1:

- Place the patient in a seated position at a table.

- Set three objects (e.g., an empty drinking glass, a set of keys, a phone) on the table in front of the patient. The objects should be placed at arm's-length distance from the patient in three separate positions, requiring the patient to reach to the left, right, and directly in front of his or her body.

- Instruct the patient to reach for each object as each one's name is called.

☞ Impairment is indicated if the patient over- or undershoots two out of the three objects. Impairment also is indicated if the patient attempts to stabilize his or her upper extremity against the table.

Procedure 2:

- Place the patient in a seated position at a table.

- Place a standard-sized drinking mug in front of the patient and give the patient a pencil.

- Instruct the patient to place the pencil in the mug. Allow the patient three attempts.

☞ Impairment is indicated if the patient over- or undershoots the mug on two out of three attempts. Impairment also is indicated if the patient attempts to stabilize his or her upper extremity against the table.

4. *Dysdiadochokinesia* is an impaired ability to perform rapid alternating movements, such as bilateral forearm pronation–supination or bilateral hand grasp–release. The patient's attempt at rapid alternating movements becomes irregular; bilateral movements cease to be simultaneous. Often one extremity lags behind the other.

- Place the patient in a seated position with elbows flexed to 90°.

- Instruct the patient to turn his or her palms up and down (forearm pronation–supination) rapidly and simultaneously (see Figure 6.5).

A variation on this procedure requires the patient to tap his or her hands against the thighs as he or she rapidly supinates and pronates the forearms (palmer surface taps thighs, dorsal surface taps thighs). The ability to rapidly reverse movements bilaterally can be assessed at any joint, for example, flexing and extending both knees simultaneously, flexing and extending the fingers simultaneously, and plantar flexion and dorsiflexion of both ankles simultaneously.

Figure 6.5 Dysdiadochokinesia

☞ Impairment is indicated if the patient experiences difficulty performing the requested movement bilaterally at the same speed and range of motion. The patient will lose the rhythm and pace of the movement; often one extremity will lag behind the other in speed and range of motion.

5. *Rebound phenomenon* is the inability to regulate the action of opposing muscle groups.

- Place the patient in a seated position with elbow flexed (see Figures 6.6 Aand B).

- Ask the patient to resist your attempt to pull his or her elbow into extension.

- Release the patient's forearm. Normally the elbow should remain in approximately the same position due to the regulation of opposing muscle groups.

Figure 6.6A Rebound Phenomenon

- A patient with cerebellar lesions will be unable to regulate his or her opposing muscle groups, and the limb will suddenly hit the torso. This occurs as a result of impaired proprioceptive feedback—the patient cannot regulate the speed and force of opposing muscle groups quickly enough to prevent the arm from hitting the torso. The rebound phenomenon can be assessed using other muscle groups, such as the shoulder abductors or flexors.

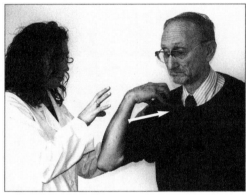

Figure 6.6B Rebound Phenomenon

☞ Impairment is indicated if the patient is unable to regulate the speed and force of the opposing muscle groups.

6. *Asthenia* is muscle weakness. A generalized asthenia commonly occurs with posterior and flocculonodular cerebellar lobe damage.

Fixation or position holding:

- Place the patient in a seated position.

- Instruct the patient to fixate or maintain a specific position, such as bilateral shoulder flexion or abduction at 90°.

☞ Impairment is indicated if the patient tires quickly and cannot maintain the position for more than 30 seconds.

Manual muscle testing (MMT): The presence of asthenia also can be detected through MMT of the major muscle groups (see Section 5; "Motor Screening").

☞ Impairment is indicated if the major muscle groups on one side of the body (the involved side) are weaker than the uninvolved side. Impairment also may also be indicated if the patient's generalized strength is significantly weaker than normal for the person's age and weight.

7. *Motor impersistence* occurs when a patient attempts to maintain both extremities in the same position but, unbeknownst to the patient, the involved extremity drifts out of its position. Motor impersistence can be tested along with asthenia.

Fixation or position holding:

- Place the patient in a seated position.

- Instruct the patient to fixate or maintain a specific position, such as bilateral shoulder flexion or abduction at 90°.

☞ Impairment is indicated if one of the patient's upper extremities (on the involved side) drifts out of its position (e.g., one of the patient's arms drops below 90° of shoulder flexion) unbeknownst to the patient.

8. *Hypotonicity* and *hyporeflexia* may occur as a result of posterior and flocculonodular cerebellar lobe damage. Hypotonicity is a decrease in muscle tone; hyporeflexia is a decrease in or absence of deep tendon reflexes.

- Place the patient in a seated position.

- Palpate the patient's major muscle groups and passively move the patient's major joints through their range of motion.

☞ Hypotonicity is indicated by a diminished resistance to passive movement. The patient's muscles will feel soft or flaccid on palpation.

- Attempt to elicit the patient's deep tendon reflexes (see "Screening for Deep Tendon Reflexes" in Section 5).

☞ Hyporeflexia is indicated by diminished or absent deep tendon reflexes.

9. *Hypertonicity,* particularly *extensor tone,* can result from anterior cerebellar lobe damage. Hypertonicity is an increase in muscle tone. *Hyperreflexia*—increased deep tendon reflexes—often will accompany hypertonicity.

- Place the patient in a seated position or supine on a mat.

- Palpate the patient's major muscle groups and passively move the patient's major joints through their range of motion.

☞ Hypertonicity is indicated by an increased resistance to passive movement. Because increased extensor tone commonly occurs as a result of anterior cerebellar lobe damage, the patient's joints will exhibit an increased resistance to passive flexion. The patient's muscles will feel rigid on palpation.

- Attempt to elicit the patient's deep tendon reflexes (see "Screening for Deep Tendon Reflexes" in Section 5).

☞ Hyperreflexia is indicated by an increase in deep tendon reflexes.

10. *Ataxia* is an umbrella term used to describe incoordinated patterns of movement that affect one's gait, posture, and upper-extremity motor control. An *ataxic gait* is characterized by a wide base of support with arms held away from the body to enhance balance. Ambulation is unsteady, and the patient appears to stagger as he or she walks; the patient also is unable to maintain a straight, direct-forward line while walking and, instead, tends to veer toward the side of the lesion. An *ataxic posture* is characterized by back-and-forth oscillations of the body while standing upright. Ataxic patterns of the upper extremities appear as up-and-down oscillating movements when the limbs are held against gravity.

- Ask the patient to stand position with eyes open. Observe the patient's upright posture and bodily sway.

- Instruct the patient to remain standing but with eyes closed. Note whether the patient's balance deteriorates without visual cues.

☞ The inability to maintain an upright standing posture without visual cues is referred to as a *positive Romberg sign*. If the patient is able to maintain balance with eyes open but not with eyes closed, proprioceptive loss is indicated.

- As the patient is standing with eyes open, gently displace his or her balance unexpectedly. Maintain close contact guard to prevent falls or injury.

- Repeat the procedure with the patient's eyes closed.

☞ Impairment is indicated if the patient easily loses balance when displaced by the therapist. If the patient is able to maintain balance with eyes open but not with eyes closed, proprioceptive loss is indicated.

- Ask the patient to stand on one foot, first with eyes open, then with eyes closed.

- Observe the patient's ability to maintain balance on one foot. Note whether balance deteriorates with eyes closed.

☞ Impairment is indicated if the patient is unable to maintain balance on one foot.

- Instruct the patient to perform the following movements:
  - Walk on a straight line taped to the floor.
  - Walk sideways and backward.

– March in place.

– Walk on heels or toes.

☞ Impairment is indicated if the patient easily loses balance while performing any of these movements.

11. *Nystagmus* is an involuntary back-and-forth movement of the eyes in a jerky, oscillating fashion when the eyes move laterally or medially to any of the visual field extremes. Nystagmus occurs because the cerebellum has connections to the extraoculomotor muscles via the medial longitudinal fasciculus. The medial longitudinal fasciculus is a brain stem tract that coordinates head and eye movements. This connection allows the cerebellum to exert influence on the tone and coordinated movements of the extraoculomotor muscles.

- Sit in front of the patient, who also is seated.

- Instruct the patient to maintain his or her head in a fixed position while he or she visually scans a moving object. The moving object can be a colored pen cap.

- Move the visual stimulus in front of the patient in the shape of an *H* and an *X*.

- Note whether nystagmus is evident as the patient's eyes move into the visual field extremes while scanning the visual stimulus. The visual field extremes are the superior and inferior quadrants and the temporal and nasal quadrants (see Figure 6.7).

☞ Cerebellar damage may be indicated if nystagmus can be elicited when the patient's eyes move into the visual field extremes.

12. *Dysarthria* or *staccato voice* involves an impairment in the motor movements of speech. The patient's use of language and grammar remain intact, but the ability to clearly enunciate words is impaired. The modulation of the motor movements involved in speech is a proprioceptive or cerebellar function. Dysarthria or staccato voice occurs because the cerebellum cannot regulate the rate and coordination of speech motor patterns. Speech is broken, with prolonged, slow, slurred syllables.

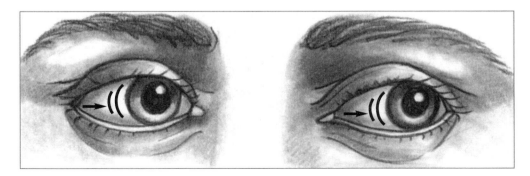

Figure 6.7 Nystagmus

■ Observe the patient's speech during the screening procedures.

■ Ask the patient to read a paragraph or to repeat phrases first spoken by the therapist:

– The quick brown fox jumped over the fence.

– A penny saved is a penny earned.

– Mary had a little lamb whose fleece was white as snow.

☛ Cerebellar damage may be indicated if the patient's speech is dysarthric or is characterized by a staccato voice.

BASAL GANGLIA SCREENING PROCEDURES

1. *Tremors at rest,* or *nonintention tremors,* are involuntary oscillating movements that occur in an extremity at rest. Resting tremors decrease or disappear with the initiation of voluntary movement. Resting tremors tend to worsen with increased emotional stress. Resting tremors often are often associated with Parkinson's disease (a basal ganglia disorder) and are characterized by a *pill-rolling movement*—a tremor in which the patient appears to be rolling a pill between his or her thumb and first two fingers. Resting tremors may also be observed at the wrist in the form of hand tremors and at the forearm in the form of pronation–supination tremors. A head tremor also may be observed.

■ Observe the patient at rest. Note whether tremors at rest can be detected (e.g., pill-rolling, forearm pronation–supination, hand tremors, head tremor).

■ If tremors are detected, ask the patient to perform an activity using the limbs in which a tremor is noted. For example, instruct the patient to perform at least one of the following tasks:

– Make a sandwich.

– Pour water from a container into a drinking glass.

– Fold a piece of paper and place it in an envelope.

☛ Impairment is indicated if tremors at rest are detected and if the tremors disappear or decrease when voluntary movement is initiated at that limb.

2. *Rigidity* is a form of hypertonicity (increased muscle tone) that is characterized by increased resistance to passive movement of a joint in all planes. Rigidity differs from spasticity in that spasticity involves increased tone in either the flexors or the extensors of a joint, but not both. Rigidity involves increased tone in both the flexors and the extensors (and the abductors and adductors) of a joint.

■ Perform passive range of motion of all the patient's joints.

☛ Two types of rigidity are signs of basal ganglia impairment. *Lead pipe rigidity* is characterized by a uniform and continuous resistance to passive movement as the extremity is moved through its range of motion (in all planes). Lead pipe rigidity is indicated if a uniform and continuous resistance to passive movement is detected when attempting to passively move an extremity through its range of motion (in all planes). *Cogwheel rigidity* is characterized by an alternate release–resistance pattern to passive movement as the extremity is moved through its range of motion (in all planes). Cogwheel rigidity can be felt as a series of brief muscle relaxations followed by quick contractions. It is a common sign of Parkinson's disease. Cogwheel rigidity is indicated if an alternate release–resistance pattern is detected when attempting to passively move an extremity in its range of motion (in all planes).

3. *Akinesia* is an inability to initiate voluntary movement; it is commonly seen in the late stages of Parkinson's disease. Generally, the patient assumes a fixed posture as a result of the inability to initiate movement. Patients report that a tremendous amount of mental concentration is required to perform the most basic motor tasks (e.g., raising one's arm to comb one's hair, bringing a fork to one's mouth to eat).

4. *Bradykinesia* is slowed or decreased movement; it is sometimes referred to as *poverty of movement*. In addition to slowed movements, the patient's ability to quickly change movements becomes delayed; switching from one motor pattern to another becomes difficult. Bradykinesia also can manifest as a lack of facial expression (or masked face), monotone speech, and reduced eye movement.

Akinesia and bradykinesia can be observed during the course of screening procedures. The therapist should note (a) whether the patient can initiate voluntary movement, (b) whether the patient's voluntary movements are abnormally slowed, and (c) whether the patient has difficulty switching from one motor pattern to another.

- Ask the patient to demonstrate different facial expressions—happiness, sadness, anger. Is the patient able to exhibit various facial expressions, or is facial expression masklike?

- Ask the patient to change the inflection of his or her voice to sound angry, sad, happy. Can the patient change the inflection of his or her voice, or is speech monotone?

- Present a comb to the patient and ask him or her to comb the hair on the right side of his or her head. Alternatively, provide the patient with hand lotion and ask him or her to rub it into the skin on the opposite arm. Observe whether the patient is able to initiate movement and how long initiation of movement takes.

- Ask the patient to comb the hair on the left side of his or her head (or to rub hand lotion into the opposite arm) to determine whether the patient can abruptly change movements. Observe if the patient is quickly able to change motor patterns, or if changing motor patterns is abnormally delayed.

☛ Impairment is indicated if the patient's facial expressions are masklike, if monotone speech is detected (meaning the patient is unable to regulate the rate and inflection of vocal tone), if the patient is unable to voluntarily initiate motor patterns, if voluntary movements are abnormally slowed, or if the patient is unable to change movement patterns in a timely manner when performing functional activities.

5. *Cunctation-festinating gait. Cunctation* means to resist movement. *Festination* means to hurry. A cunctation-festinating gait is characterized by difficulty both initiating and stopping walking. Once the patient is able to begin walking, the movement patterns for walking become hurried. The patient appears to quickly shuffle. Reciprocal arm swing often is absent; however, sometimes a patient may demonstrate an exaggerated arm swing to enhance propulsion of movement. The patient also demonstrates an inability to stop walking once started and often bumps into walls or furniture. Changing directions while walking is a difficult task—patients often are unable to circumvent obstacles in their path once they have initiated a particular walking direction. Balance and equilibrium responses often are decreased or absent. A cunctation-festinating gait is commonly seen in Parkinson's disease.

- Ask the patient to walk along a straight line (approximately 10 feet, if possible).

- Observe how long it takes the patient to initiate walking. Once the patient begins to walk, observe whether reciprocal arm swing is present. Note whether the patient's gait is hurried.

- While the patient is walking, ask him or her to alter the speed and direction: "Walk faster. Turn to the right and keep walking. Walk slower. Turn to the left and keep walking. March in place."

- Ask the patient to suddenly stop walking. Observe whether the patient is easily able to stop walking or has difficulty.

- When standing still, gently displace the patient's balance. Maintain close contact guard with the patient. Observe whether the patient easily loses balance. Note whether protective responses are present, decreased, or absent.

☞ Impairment is indicated if the patient demonstrates difficulty starting and stopping walking, if the patient's gait appears hurried and is characterized by small shuffling steps, if reciprocal arm swing is absent or exaggerated, if the patient has difficulty altering the speed and direction of walking, or if the patient easily loses balance when gently displaced and exhibits impaired or absent protective responses.

6. *Chorea,* or *choreiform movements,* are sudden, rapid, involuntary jerky movements that primarily involve the face and extremities. Chorea is associated with Huntington's disease—a degenerative disorder involving the caudate and putamen. Shoulder shrugs, hip movements, crossing and uncrossing one's legs, facial grimaces, and tongue protrusions are signs of chorea. The presence of choreiform movements can be detected during screening procedures by noting whether the patient presents with any of these symptoms.

7. *Athetosis,* or *athetoid movements,* are slow, flailing, twisting movements that are wormlike in quality. Athetosis often presents in combination with spasticity or hypotonicity and is believed to result from damage to the caudate and/or putamen. Athetoid movements commonly involve the neck, face, trunk, and extremities. Athetosis is a clinical feature of cerebral palsy. Athetoid movements can be detected during screening procedures by noting whether the patient presents with any of these symptoms.

8. *Hemiballismus* is characterized by violent thrashing movements of the extremities on one side of the body (the side that is contralateral to the lesioned basal ganglia). It results from a lesion to the subthalamus and caudate. Hemiballismus can be detected during screening procedures by noting whether the patient presents with these unilateral symptoms.

9. *Dystonia* is a movement disorder resulting from increased muscle tone causing distorted, twisted postures of the trunk and proximal extremities. The sustained muscle contractions of dystonia—referred to as *torsion spasms*—can last from seconds to hours. Common torsion spasms include *blepharospasm, torticolis,* and *truncal dystonia* (see glossary below). Dystonia can be detected during screening procedures by noting whether the patient presents with any of these torsion spasms.

10. *Tics* are repetitive, brief, rapid, involuntary movements involving single muscles or multiple muscle groups. Tics are caused by an increased sensitivity to dopamine in the basal ganglia. With increased sensitivity to dopamine, the caudate, which normally acts like a brake on extraneous movements, cannot suppress movements like tics. A tic can involve a brief isolated movement, such as eye blinks, head jerks, or shoulder shrugs. They may involve a variety of sounds, such as throat clearing, grunting, or the repetition of words. Tics can be detected during screening procedures by noting whether the patient presents with any of these symptoms.

Glossary of Torsion Dystonia Terms

Torsion dystonia involving the eye	
Blepharospasm	Eyes are involuntarily kept closed
Oculogyric crises	Attacks of forced deviation of gaze, often associated with a surge of Parkinsonism, catatonia, tics, and obsessiveness
Opthalomoplegia	Paralysis of gaze
Torsion dystonia involving the throat, jaw, lips, or tongue	
Oromandibular dystonia	Involuntarily opening and closing of the jaw, retraction, or puckering the lips
Lingual dystonia	Repetitive protrusion of the tongue or upward deflection of the tongue toward the hard palate
Laryngeal dystonia	Speech is tight, constricted, and forced; smooth flow of speech is lost
Pharyngeal dystonia	Associated with dysphagia, dysphonia (hoarseness), and dysarthria (slurred words)
Torsion dystonia involving the neck	
Torticollis	Dystonic contractions of the neck muscles
Retrocollis	Neck forced backward into hyperextension
Anterocollis	Neck forced forward into hyperflexion
Laterocollis	Neck forced laterally
Torsion dystonia involving the trunk	
Truncal dystonia	Manifests as a lordosis, scoliosis, tortipelvis, or opisthotonos (forced flexion of the head on the chest)
Torsion dystonia involving the legs	
Crural dystonia	Dystonic movement of the legs
Torsion dystonia of single or multiple body parts	
Focal dystonia	Dystonia in a single body part
Multifocal dystonia	Dystonia of more than one body part
Hemidystonia	Involvement of limbs on one side of the body (due to a space occupying lesion or infarction)
Generalized dystonia	Dystonia in a leg, the trunk, and another body part; or in both legs and the trunk

Note. As with most basal ganglia movement disorders, the abnormal movements usually disappear during sleep and worsen, with anxiety, fatigue, temperature changes, and pain.

RED FLAGS
SIGNS AND SYMPTOMS OF CEREBELLAR DYSFUNCTION

A red flag is an indicator or symptom of dysfunction. If a patient is observed displaying any of the following red flags, it is likely that the patient possesses cerebellar dysfunction.

- **INTENTION TREMORS** occur during voluntary movement of a limb and tend to increase as the limb nears its intended goal. For example, the patient may experience increased tremors as he or she attempts to use his or her hand to bring a spoon to the mouth. Intention tremors tend to diminish or stop when the patient's limbs are at rest.

- **MOVEMENT DECOMPOSITION**, or dyssynergia is characterized by movements that are broken up into their component parts rather than occurring as a smooth, single movement. For example, when the patient is instructed to touch his or her fingertip to the nose, he or she may first flex the elbow, then flex the shoulder, and finally adjust the position of the wrist and fingers.

- **DYSMETRIA** is an inability to judge the distance and range of a movement. It is characterized by overshooting (past pointing) or undershooting one's reach for a target object.

- **DYSDIADOCHOKINESIA** is an impaired ability to perform rapid alternating movements, such as bilateral forearm pronation–supination or bilateral hand grasp–release. The patient's attempt at rapid alternating movements becomes irregular; bilateral movements cease to be simultaneous. Often, one extremity lags behind the other.

- **REBOUND PHENOMENON** is the inability to regulate the action of opposing muscle groups. For example, the patient is asked to resist the therapist's attempt to pull the patient's elbow out of flexion and into extension. The therapist then releases the patient's forearm. A patient with cerebellar lesions will be unable to regulate the opposing muscle groups, and the limb will suddenly hit his or her torso. This occurs as a result of impaired proprioceptive feedback; the patient cannot regulate the speed and force of opposing muscle groups quickly enough to prevent the arm from hitting his or her torso.

- **ASTHENIA** is muscle weakness. A generalized asthenia commonly occurs with posterior and flocculonodular cerebellar lobe damage.

- **MOTOR IMPERSISTENCE** occurs when a patient attempts to maintain both upper extremities in the same position but, unbeknownst to the patient, the involved limb drifts out of its position.

- **HYPOTONICITY AND HYPOREFLEXIA** may occur as a result of posterior and flocculonodular cerebellar lobe damage. *Hypotonicity* is a decrease in muscle tone; *hyporeflexia* is a decrease in or absence of deep tendon reflexes.

- **HYPERTONICITY**, particularly *extensor tone*, can result from anterior cerebellar lobe damage. *Hypertonicity* is an increase in muscle tone.

- **HYPERREFLEXIA**, increased deep tendon reflexes, often will accompany hypertonicity.

- **ATAXIA** is an umbrella term used to describe incoordinated patterns of movement that affect one's gait, posture, and upper-extremity motor control. An *ataxic gait* is characterized by a wide base of support with arms held away from the body to enhance balance. Ambulation is unsteady, and the patient appears to stagger as he or she walks; the patient also is unable to maintain a straight, direct-forward line while walking and, instead, tends to veer toward the side of the lesion. An *ataxic posture* is characterized by back-and-forth oscillations of the body while standing upright. Ataxic patterns of the upper extremities appear as up-and-down oscillating movements when the limbs are held against gravity.

- **NYSTAGMUS** is an involuntary back-and-forth movement of the eyes in a jerky, oscillating fashion when the eyes move laterally or medially to any of the visual field extremes.

- **DYSARTHRIA,** or staccato voice involves an impairment in the motor movements of speech. The patient's use of language and grammar remain intact, but the ability to clearly enunciate words is impaired. The modulation of the motor movements involved in speech is a proprioceptive or cerebellar function. Dysarthria and staccato voice occur because the cerebellum cannot regulate the rate and coordination of speech motor patterns. Speech is broken, with prolonged, slow, slurred, syllables.

RED FLAGS
SIGNS AND SYMPTOMS OF BASAL GANGLIA DISORDERS

A red flag is an indicator or symptom of dysfunction. If a patient is observed displaying any of the following red flags, it is likely that the patient possesses basal ganglia disorders.

- **TREMORS AT REST,** or nonintention tremors, are involuntary oscillating movements that occur in an extremity at rest. Resting tremors decrease or disappear with the initiation of voluntary movement. Resting tremors tend to worsen with increased emotional stress.

- **RIGIDITY** is a form of hypertonicity (increased muscle tone) that is characterized by increased resistance to passive movement of a joint in all planes. Rigidity differs from spasticity in that spasticity involves increased tone in either the flexors or the extensors of a joint, but not both. Rigidity involves increased tone in both the flexors and the extensors (and the abductors and adductors) of a joint.

- **AKINESIA** is an inability to initiate voluntary movement.

- **BRADYKINESIA** is slowed or decreased movement; it is sometimes referred to as *poverty of movement.* In addition to slowed movements, the patient's ability to quickly change movements becomes delayed; switching from one motor pattern to another becomes difficult. Bradykinesia also can manifest as a lack of facial expression (or masked face), monotone speech, and reduced eye movement.

- **CUNCTATION-FESTINATING GAIT.** *Cunctation* means to resist movement. *Festination* means to hurry. A cunctation-festinating gait is characterized by difficulty both initiating and stopping walking. Once the patient is able to begin walking, the movement patterns for walking become hurried. The patient appears to quickly shuffle. Reciprocal arm swing often is absent. The patient also demonstrates an inability to stop walking once started and often bumps into walls or furniture. Changing directions while walking is a difficult task—patients often are unable to circumvent obstacles in their path once they have initiated a particular walking direction. Balance and equilibrium responses often are decreased or absent.

■ **CHOREA,** or choreiform movements, are sudden, rapid, involuntary jerky movements that primarily involve the face and extremities. Shoulder shrugs, hip movements, crossing and uncrossing one's legs, facial grimaces, and tongue protrusions are signs of chorea.

■ **ATHETOSIS,** or athetoid movements, are slow, flailing, twisting movements that are wormlike in quality. Athetosis often presents in combination with spasticity or hypotonicity. Athetoid movements commonly involve the neck, face, trunk, and extremities.

■ **HEMIBALLISMUS** is characterized by violent thrashing movements of the extremities on one side of the body (the side that is contralateral to the lesioned basal ganglia).

■ **DYSTONIA** is a movement disorder resulting from increased muscle tone causing distorted, twisted postures of the trunk and proximal extremities. The sustained muscle contractions of dystonia—referred to as *torsion spasms*—can last from seconds to hours.

■ **TICS** are repetitive, brief, rapid, involuntary movements involving single muscles or multiple muscle groups. A tic can involve a brief isolated movement, such as eye blinks, head jerks, or shoulder shrugs. Tics also may involve a variety of sounds, such as throat clearing, grunting, or the repetition of words.

AVAILABLE IN-DEPTH ASSESSMENTS

If balance and coordination screening detects impairment, the following in-depth assessments are available:

Activities-Specific Balance Confidence Scale (ABC)
L. E. Powell, A. M. Myers
Adults with balance deficits
Powell, L. E., & Myers, A. M. (1995). The Activities-Specific Balance Confidence (ABC) Scale. *Journal of Gerontology, 50A*(1), M28–M34.

Assessment of Motor and Process Skills (AMPS)
A. G. Fisher
Children and adults with developmental, neurologic, and musculoskeletal disorders
Anne G. Fischer, ScD, OTR/L, FAOTA
AMPS Project
Occupational Therapy Building
Colorado State University
Fort Collins, CO 80523

Berg Balance Scale (Berg)
K. O. Berg, S. L. Wood-Dauphinee, J. I. Williams, B. Maki
Adults with balance deficits
Berg, K. O., Wood-Dauphinee, S. L., Williams, J. I., & Maki, B. (1992). Measuring balance in the elderly: Validation of an instrument. *Canadian Journal of Public Health, 83,* S7–S11.

Box and Block Test of Manual Dexterity
V. Mathiowetz, G. Volland, N., Kashman, K. Weber
Adults with fine motor deficits secondary to neurologic and/or musculoskeletal impairment
Mathiowetz, V., Volland, G., Kashman, N., & Weber, K. (1985). Adult norms for the Box and Block Test of Manual Dexterity. *American Journal of Occupational Therapy, 39,* 386–391.

Clinical Test of Sensory Interaction in Balance (CTSIB)
M. Watson
Adults with balance deficits

Watson, M. (1992). Clinical Test of Sensory Interaction in Balance. *Physiotherapy Theory and Practice, 8,* 176–178.

Falls Efficacy Scale (FES)
M. E. Tinnetti, D. Richman, L. Powell
Elderly persons with balance deficits
Tinnetti, M. E., Richman, D., & Powell, L. (1990). Falls efficacy as a measure of fear of falling. *Journal of Gerontology, 45,* 239–243.

Functional Reach Test (FRT)
P. W. Duncan, D. K. Weiner, J. Chandler, S. A. Studenski
Adults with balance deficits secondary to neurologic damage
Duncan, P. W., Weiner, D. K., Chandler, J., & Studenski, S. A. (1990). Functional reach: A new clinical measure of balance. *Journal of Gerontology, 45,* 192–197.

Jebsen Hand Function Test
R. H. Jebsen, N. Taylor, R. B. Trieschmann, M. J. Trotter, L. A. Howard
Children and adults with neurologic and/or musculoskeletal impairment
Sammons Preston Inc.
PO Box 50710
Bolingbrook, IL 60440-5071
(800) 323-5547

Minnesota Rate of Manipulation Tests (MRMT) and Minnesota Manual Dexterity Test (MMDT)
University of Minnesota Employment
   Stabilization Research Institute
Adults with neurologic and/or musculoskeletal impairment
American Guidance Services, Inc.
Publishers' Bldg.
Circle Pines, MN 55014
(800) 328-2560

Morse Fall Scale (MFS)
J. M. Morse
Adults with balance deficits
Janice M. Morse
Pennsylvania State University
School of Nursing

201 Health and Human Development East
University Park, PA 16802-6508

Nine-Hole Peg Test
V. Mathiowetz, K. Weber, N. Kashman, G. Volland
Adults with fine motor coordination deficits secondary to neurologic and/or musculoskeletal impairment
Mathiowetz, V., Weber, K., Kashman, N., & Volland, G. (1985). Adult norms for the Nine-Hole Peg Test of finger dexterity. *Occupational Therapy Journal of Research, 5,* 24–38.

O'Connor Finger and Tweezer Dexterity Tests
Adults with fine motor coordination deficits secondary to neurologic and/or musculoskeletal impairment
Sammons Preston Inc.
PO Box 5071
Bolingbrook, IL 60440-5071
(800) 323-5547

Purdue Pegboard
J. Tiffin
Adults with fine motor and eye–hand coordination deficits secondary to neurologic and/or musculoskeletal impairment
Lafayette Instrument Company
3700 Sagamore Pkwy. North
PO Box 5729
Lafayette, IN 47903
(800) 428-7545

Roeder Manipulative Aptitude Test
W. S. Roeder
Adolescents and adults with fine motor and eye–hand coordination deficits secondary to neurologic and/or musculoskeletal impairment
Lafayette Instrument Company
3700 Sagamore Pkwy. North
PO Box 5729
Lafayette, IN 47903
(800) 428-7545

Physical Performance Test (PPT)
D. B. Reuben, A. L. Siu

Elderly persons with balance deficits
Reuben, D. B., & Siu, A. L. (1990). An objective measure of physical function of elderly outpatients: The Physical Performance Test. *Journal of the American Geriatric Society, 38,* 1105–1112.

Timed Get Up and Go (TGUG)
D. Podsiadlo, S. Richardson
Adults with balance deficits
Podsiadlo, D., & Richardson, S. (1991). The Timed Get Up and Go: A test of basic functional mobility for frail elderly persons. *Journal of the American Geriatric Society, 39,* 142–148.

Tinnetti Balance Test of the Performance-Oriented Assessment of Mobility Problems (Tinnetti)
M. E. Tinnetti
Adults with balance deficits
Tinnetti, M. E. (1986). Performance-oriented assessment of mobility in elderly patients.

Journal of the American Geriatric Society, 34, 119–126.

For further information about the these assessments, refer to the following resources:

Asher, I. E. (1996). *Occupational therapy assessment tools: An annotated index* (2nd ed.). Bethesda, MD: American Occupational Therapy Association.
Impara, J. C., Plake, B. S., & Murphy L. L. (Eds.). (1998). *The thirteenth mental measurements yearbook.* Lincoln: University of Nebraska Press.
Murphy, L. L., Impara, J. C., & Plake, B. S. (Eds.). (1999). *Tests in print: An index to tests, test reviews, and the literature on specific tests* (Vols. 1 & 2). Lincoln: University of Nebraska Press.
Plake, B. S., Impara, J. C., & Murphy, L. L. (Eds.). (1999). The supplement to The Thirteenth Mental Measurements Yearbook. Lincoln: University of Nebraska Press.

CEREBELLAR DISORDER SCREENING

Patient Name _____ Date _____

Sign/Symptom	Present	Absent	Comments
Intention tremors			
Movement decomposition (Dyssynergia)			
Dysmetria			
Dysdiadochokinesia			
Rebound phenomenon			
Asthenia			
Hypotonicity			
Hyporeflexia			
Hypertonicity			
Hyperreflexia			
Ataxic gait			
Ataxic posture			
Nystagmus			
Dysarthria			

BASAL GANGLIA DISORDER SCREENING

Patient Name _____   Date _____

Tremors at Rest   ☐ Present   ☐ Absent   Indicate in which body part(s) resting tremors are detected: _____

_____

Rigidity   ☐ Present   ☐ Absent   ☐ Lead pipe   ☐ Cogwheel   Indicate at which joints rigidity is detected: _____

_____

Akinesia/Bradykinesia

Is patient able to initiate voluntary movement?   ☐ Yes   ☐ No

Is voluntary movement abnormally slow?   ☐ Yes   ☐ No

Does patient have difficulty switching from one motor pattern to another?   ☐ Yes   ☐ No

Facial expression: ☐ Masklike   ☐ Normal

Speech tonal inflection: ☐ Monotone   ☐ Normal

Cunctation-Festinating Gait

Does patient have difficulty initiating walking?   ☐ Yes   ☐ No

Does patient have difficulty stopping walking?   ☐ Yes   ☐ No

Does patient exhibit a hurried, shuffling gait?   ☐ Yes   ☐ No

Reciprocal arm swing: ☐ Present   ☐ Absent

Is patient able to easily change speed and direction while walking?   ☐ Yes   ☐ No

Balance and equilibrium responses: ☐ Intact   ☐ Decreased   ☐ Absent

Chorea

Does patient exhibit sudden, rapid, involuntary   ☐ Yes   ☐ No   Indicate in which body part(s) chorea is
jerky movements of the face and/or extremities?   detected: _____

_____

Athetosis

Does patient exhibit slow, flailing, twisting   ☐ Yes   ☐ No   Indicate in which body part(s) athetosis is
movements of the neck, face, trunk, and/or   detected: _____
extremities?

_____

Hemiballismus

Does patient exhibit violent thrashing of extremities   ☐ Yes   ☐ No   Indicate in which body part(s)
on one side of the body?   hemiballismus is detected: _____

_____

Dystonia

Does patient exhibit increased muscle tone causing   ☐ Yes   ☐ No   Indicate which body parts are involved in
distorted, twisted postures of the trunk and   the dystonic posture: _____
extremities?

_____

Tics

Does patient exhibit repetitive, brief, rapid   ☐ Yes   ☐ No   Describe the type of tic detected: _____
involuntary movements or sounds?

_____

Cranial Nerve Function
SCREENING

FUNCTIONAL IMPLICATIONS OF CRANIAL NERVE DAMAGE

The *cranial nerves* are 12 pairs of *peripheral nerves* that exit the brain between the midbrain and the medulla. The cranial nerves are considered to be part of the peripheral nervous system. The cell bodies of the cranial nerves are located in the brain stem; thus, cranial nerve nuclei are considered to be part of the central nervous system (CNS). The cranial nerves carry sensory and motor information to and from the receptors of the head, face, and neck. When neurologic impairment occurs to a cranial nerve, *ipsilateral symptoms* result.

Several of the cranial nerves—1, 2, 7, 8, and 9—mediate the function of the *special senses:* vision, audition, olfaction, and gustation. Damage to these cranial nerves can impair sight, hearing, smell, and taste. When vision in one eye is lost, the patient may not attend to objects in the lost visual field. Safety risks—from a failure to observe objects in the lost visual field—become problematic. For example, while crossing the street, a patient may fail to observe an oncoming car turning in his or her lost visual field. Safety risks also occur if hearing is lost or impaired; the patient may fail to identify by sound the direction of oncoming traffic. Similarly, if smell is lost or impaired, the patient may fail to detect food burning on the stove that, left unattended, could ignite a fire. Often, if smell is lost, taste also becomes impaired. If both taste and smell are lost or impaired, the patient may fail to identify spoiled food and become ill from its ingestion.

The *extraoculomotor cranial nerves*—3, 4, and 6—mediate eyeball movements. When one or more of these cranial nerves are damaged, vision may remain but be impaired by a strabismus, nystagmus, diplopia, or lost pupillary reflex. As a result, the patient may have difficulty negotiating the environment. For example, a patient who displays a subtle vertical, medial strabismus may have difficulty walking down steps. If the strabismus is not identified and treated, the patient is at an increased risk of injury when negotiating environments in which steps are present. A patient who has diplopia, or double vision, may fail to grasp the handle of a pot accurately and, consequently, scald him- or herself by spilling the boiled contents of the pot on his or her body.

Cranial nerve 8, the *vestibulocochlear nerve,* mediates the function of equilibrium. When damage occurs to this cranial nerve, the patient experiences decreased balance and protective responses. A decrease in balance and protective responses places the patient at a greater safety risk for falls. Any environment that requires transitional movements (e.g., stepping down from a curb, getting in and out of a car, rising from a seat and walking down the aisle of a moving bus) poses a safety risk.

Several of the cranial nerves—5, 7, 9, 10, 11, and 12—play important roles in the functions of *eating* and *swallowing*. When one or more of these cranial nerves are damaged, swallowing difficulty, choking, or food aspiration can occur.

CRANIAL NERVE SCREENING PROCEDURES

CRANIAL NERVE 1: OLFACTORY NERVE

The function of the olfactory nerve is smell.

- Screen one olfactory nerve at a time.

- Occlude the patient's vision.

- Block the patient's nostril on the opposite side being screened.

- Present one odor at a time. Present at least four familiar scents, such as peppermint, cinnamon, coffee, and vanilla. Ammonia and acetic acid should be avoided because their pungency can stimulate nerve endings in the mucous membranes, giving a false-positive result. The patient may think that he or she is smelling the odor when in fact he or she is tasting the odor's dissolving vapors.

- Provide the patient with a verbal choice of specific odors if he or she has difficulty identifying the odors but can easily smell them or if the patient has word-finding difficulties.

☞ Olfactory nerve damage is indicated if the patient has difficulty smelling and identifying three out of the four presented odors.

Loss of smell is called *anosmia*. It often results from head injury. When the patient presents with loss of smell, loss of taste often occurs as well.

CRANIAL NERVE 2: OPTIC NERVE

The function of the optic nerve is visual acuity or the accuracy of sight, not the interpretation of visual information. When screening optic nerve function, three kinds of data should be collected, or the examination is not considered complete:

1. Results from a visual acuity test (Snellen Eye Chart)

2. Results from a visual field test

3. Results from a *fundoscopic exam* (examination of the optic nerve fibers as they leave the retina).

Generally, therapists perform visual acuity and visual field tests. Ophthalmologists perform fundoscopic exams.

Screening visual acuity. Visual acuity refers to the ability to perceive visual detail. The *Snellen Chart* is a standardized instrument for measuring visual acuity.

- Seat the patient at a distance of 20 feet away from the chart (see Figure 2.9).

- Screen one eye at a time (monocular vision).

- Screen both eyes together (binocular vision). If the patient normally wears corrective lenses, screen the patient with glasses on.

- Ask the patient to read the top line from left to right, if possible.

- Ask the patient to read each successive line below the top line.

- Note the last line on which the patient identified more than half the letters correctly.

- Document the patient's score for that eye as the distance at which the patient could read the letters (20 feet), over the number located on the chart next to that line. Also record the number of letters the patient missed in that line. For example, a score of 20/30-1 means that the patient could read at a distance of 20 feet all but one of the letters in the line that a normally functioning eye can read at 30 feet.

☛ If the patient scored 20/40 or poorer, a referral to an ophthalmologist should be made.

Screening visual fields (or confrontation testing). The eye's visual field is the area of peripheral vision when the eye is focusing straight ahead. The visual field is screened by mapping the point at which the patient is first able to identify a visual stimulus along each peripheral perimeter (see Figure 2.12). Visual fields are screened in the vertical and horizontal (or temporal) planes and in the four quadrants: superior, inferior, right, and left. When the patient is looking straight ahead, an object located directly in front of the patient is considered to be at 0°. An object located directly over the patient's head is at 90°. An object located directly out from the patient's ear is also at 90°.

- Screen one eye at a time.

- Occlude vision in the opposite eye.

- Instruct the patient to look directly ahead (at the therapist's nose). As the patient looks directly ahead, move the visual stimulus in the horizontal plane from the 90° position to the 0° position. Instruct the patient to verbally indicate when he or she first sees the visual stimulus.

- Document the position along the plane at which the patient first indicated vision of the stimulus.

- Repeat this procedure in the vertical plane.

☞ A visual field deficit may be present if the patient scored more than 5° below the above norms.

Note: Normal peripheral vertical vision is 45°. Normal horizontal/temporal peripheral vision is 85°.

If one of the peripheral fields is abnormally small, there may be a *field deficit* present. The patient should be referred for further testing by a specialist. A *blind spot* inside the visual field suggests retinal damage. *Unilateral blindness* (in one eye, but not the other) suggests an optic nerve lesion. *Bilateral field deficits* in the temporal or nasal fields of both eyes suggests an optic tract lesion. Such field deficits are called *contralateral homonymous hemianopsias*.

CRANIAL NERVE 3: OCULOMOTOR NERVE

The oculomotor nerve is one of the extraoculomotor cranial nerves, along with cranial nerves 4 and 6. The extraoculomotor nerves are important for the coordinated movements of the head and eyes. Cranial nerve 3 is responsible for the eyeball movements up, down, medially, and laterally. It innervates the levator palpebrae muscle for eyelid opening. Cranial nerve 3 mediates the *pupillary reflex* (the pupil of the eye constricts when light is shined into it) and *consensuality* (occurs when the pupil of the eye that is not receiving light constricts at the same time as the pupil of the eye that is receiving light). Consensuality occurs because the optic nerve of the eye that is being stimulated carries the light stimulus message to the brain. Both oculomotor nerves are then stimulated, resulting in constriction of the pupils in both eyes. Cranial nerve 3 also mediates *convergence* (occurs when the pupils move medially when viewing an object at close range).

Screening pupillary constriction.

- Place the patient in a seated position.

- Darken the room to enlarge the patient's pupils. Instruct the patient to look straight ahead.

- Shine the light from a penlight directly into the pupil of the eye being screened (see Figure 2.1). Note whether the pupil constricts quickly or sluggishly.

- Repeat the procedure for the other eye.

- Shine the light into both eyes simultaneously so that the same amount of light reaches both eyes.

- Turn the light on and off. Note whether both pupils dilate and constrict together, simultaneously, and to the same degree.

☞ Impairment is indicated if the pupil of one or both eyes constricted sluggishly when exposed to light. Impairment also is indicated if both pupils did not dilate and constrict simultaneously and to the same degree when exposed to light.

Screening consensual light reflex (consensuality).

- Place the patient in a seated position.

- Darken the room.

- Place one hand on the patient's nose between the eyes.

- Shine a penlight into one eye and observe the pupillary reflex in the other eye (both eyes should constrict).

- Document whether the pupil of the unstimulated eye does not constrict at the same time as the pupil of the other eye.

☞ Impairment is indicated if the pupil of the unstimulated eye did not constrict at the same time as the pupil of the other eye.

Screening convergence.

- Place the patient in a seated position.

- Hold a pencil (or similar visual stimulus) in front of the patient (approximately 18 inches away).

- Instruct the patient to visually fixate on the pencil.

- Begin to move the pencil in a straight line toward the patient's nose (see Figure 2.11).

- Instruct the patient to keep looking at the pencil as it is moved closer to his or her nose.

- Observe for symmetrical medial eye movement.

- Record the distance at which the convergence was broken (when the patient could no longer medially invert both eyeballs).

- Note also whether the pupils constricted as both eyes converged. In the normally functioning eyes, the pupils constrict as the eyes converge. Convergence is normally broken at approximately 3 to 4 inches from the patient's nose.

☞ Impairment is indicated if the pupils did not constrict as the eyes converged. Impairment also is indicated if convergence was broken at a distance greater than 3 to 4 inches from the patient's nose.

CRANIAL NERVES 3, 4, AND 6: OCULOMOTOR, TROCHLEAR, AND ABDUCENS NERVES

Screening the function of the eye muscles. When screening the function of the eye muscles, cranial nerves 3, 4, and 6 are screened together.

■ Screen one eye at a time.

■ Occlude the eye not being screened.

■ Instruct the patient to maintain his or her head in a fixed position while visually scanning a moving stimulus (the moving stimulus can be a colored pen cap).

■ Move the visual stimulus in the shape of an *H* and *X*, allowing you to determine whether the eyeball muscles are functioning adequately (see Figures 2.6 and 2.7).

  – The oculomotor cranial nerve allows the eye to move up, down, medially, and laterally.

  – The trochlear cranial nerve allows the eye to move downward and laterally.

  – The abducens cranial nerve allows the eye to move laterally.

■ Note each time the smoothness of the movement and the distance the eye moves as it follows the visual stimulus. Document any abnormality in the smoothness or distance of movement.

■ Repeat using both eyes.

☛ Impairment is indicated if the eyes did not follow the visual stimulus smoothly, that is, if the eyes jerked back and forth while scanning the visual stimulus. Impairment also is indicated if the eyes could not scan the entire distance up, down, medially, and laterally.

Screening for the presence of strabismus. Strabismus is a condition in which both eyeballs do not sit symmetrically in their orbital sockets. Strabismus may result from weakness in eyeball muscles or from an abnormality in the innervation of an eyeball muscle. If an eyeball muscle is abnormally innervated, its opposing muscle will pull the eyeball in the opposite direction. *Lateral strabismus* (external strabismus or exotropia) occurs when the oculomotor nerve cannot innervate the medial rectus muscle. The eyeball deviates outward because the lateral rectus muscle is working unopposed (see Figure 2.3). *Vertical, medial strabismus* occurs when the trochlear nerve is not innervating the superior oblique muscle. The medial rectus and the superior rectus muscles are working unopposed to pull the eyeball up and medially (see Figure 7.1). *Medial strabismus* (internal strabismus or esotropia) occurs when the abducens nerve is not innervating the lateral rectus muscle. The medial rectus muscle works unopposed to pull the eyeball medially (see Figure 2.2).

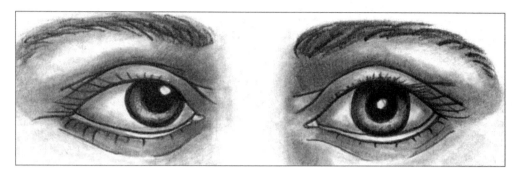

Figure 7.1 Vertical Medial Strabismus

- Darken the room.

- Shine a penlight at the bridge of the patient's nose and observe for symmetrical corneal reflection.

☛ Strabismus is indicated if the patient's corneal reflection is asymmetrical. Sometimes a strabismus is not visible to the examiner but exists to a sufficient degree to cause diplopia, or double vision.

Screening for diplopia (double vision).

- Seat the patient so that he or she faces a blank wall.

- Instruct the patient to focus on a pencil that is held approximately 18 inches in front of his or her eyes. Stand behind the patient.

- Position the pencil so that it is vertical, then horizontal, and finally diagonal. For each pencil position, instruct the patient to verbally indicate whether the image is double or single.

- Repeat the procedure holding the pencil in each of the four quadrants: superior, inferior, right, and left.

- Document the quadrant and position of the pencil for which the patient reports a double image.

- If the patient reported a double image for any of the screening conditions, repeat the procedure for each eye individually while occluding vision in the opposite eye.

☛ Impairment is indicated if the patient reports double vision for any of the screening conditions.

Screening for the presence of nystagmus. Nystagmus is a condition in which the eyeballs involuntary move back and forth in a quick, jerky, oscillating fashion when the eyes move (a) laterally or medially to either the temporal of nasal field extremes or (b) superiorly and inferiorly in the visual field extremes (see Figure 6.7). Nystagmus can be normally elicited in an intact CNS using rotational or temperature stimulation of the semicircular canals of the inner ears. Pathological nystagmus is a sign of CNS abnormality and can occur with or without external stimulation. One CNS abnormality that causes nystagmus is impairment of the extraoculo-motor cranial nerves 3, 4, and 6.

- Place the patient in a seated position.

- Instruct the patient to maintain his or her head in a fixed position while scanning a visual stimulus (the visual stimulus could be a colored pen cap).

- Move the pen cap in a letter *H* and *X* to determine whether nystagmus is elicited when the eyes move in the temporal and nasal field extremes (in the four quadrants: superior, inferior, right, and left).

☛ Impairment is indicated if nystagmus is elicited when the eyes move in any of the visual field extremes.

If nystagmus is present, determine the following:

- Is the nystagmus present at all times or only when the eye moves in certain directions?

- Are the oscillating movements of nystagmus horizontal, vertical, or rotary?

- Is the nystagmus present only when the eye changes its direction of gaze or also when the eye remains in a stationary position?

- Is the nystagmus present in one eye or both eyes?

CRANIAL NERVE 5: TRIGEMINAL NERVE

The trigeminal nerve has both sensory and motor functions. The sensory half of the trigeminal nerve mediates sensation of the face, head, cornea of eye, and inner oral cavity. The sensory half of the trigeminal nerve is responsible for the *corneal reflex* (when the cornea is touched, the eyelids close). The motor half of the trigeminal nerve innervates the jaw muscles that control chewing (or mastication). The motor half of the trigeminal nerve also is responsible for the *masseter reflex* (when the masseter M is lightly tapped with a reflex hammer, the masseter contracts).

Screening the sensory half of the trigeminal nerve. Evaluate the patient's sensory abilities on the face, head, and inner oral cavity.

- Place the patient in a seated position.

- Occlude vision.

- Use a cotton swab to stroke the patient's forehead, cheek, jaw, and chin.

- Instruct the patient to verbally identify the facial area being touched. If any areas of anesthesia are detected, map the area and document all results.

- Apply the cotton swab to the unaffected side first. Then proceed to the involved side and note any asymmetry in the patient's sensory abilities. If areas of anesthesia are detected, repeat the procedure using the bare wooden end of a cotton swab applicator to screen the patient's ability to sense sharp items.

- Apply the wooden applicator to the unaffected side first, then proceed to the involved side, noting any asymmetry in the patient's sensory abilities. If sensory deficits are found for sharp items, repeat the procedure using hot and cold stimuli.

- Apply test tubes of hot and cold water to different areas of the face in a random, alternating sequence.

- Touch the patient's cornea lightly with a cotton swab to check for corneal reflex function. The patient's eye should blink when the cornea is touched. Again, screen the unaffected side first, followed by the affected side.

☛ Impairment is indicated if areas of anesthesia are detected for light touch, sharp touch, and hot-and-cold touch. Impairment also is indicated if the corneal reflex is absent or sluggish.

Screening the motor half of the trigeminal nerve.

- Place the patient in a seated position.

- Occlude vision.

- Ask the patient to open his or her mouth. Check for deviation of the jaw to the affected side (see Figure 7.2).

- Check for asymmetry of the size of the mouth opening. If the trigeminal N is impaired, the patient will likely exhibit a decreased ability to open the mouth on the affected side.

- Ask the patient to move his or her jaw from side to side. Check for asymmetry of jaw movement.

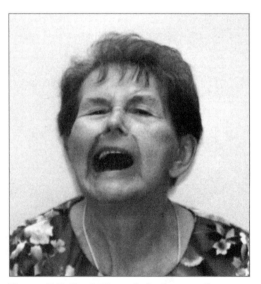

Figure 7.2 Deviation of the Jaw to the Affected Side: Screening Motor Half of Trigeminal Nerve

- Instruct the patient to bite down on a tongue depressor. Ask the patient to resist your attempts to pull the tongue depressor out. Check for asymmetry between right and left jaw strength.

☞ Impairment is indicated if the patient presents with jaw deviation to the affected side, decreased ability to open the mouth on the affected side, and decreased jaw strength on the affected side.

CRANIAL NERVE 7: FACIAL NERVE

The facial nerve has both sensory and motor functions. The sensory portion of the facial nerve innervates the taste receptors on the anterior tongue. The motor portion of the facial nerve innervates the muscles of facial expression, the muscles for eyelid closing, and the stapedius muscle (which controls the stapes of the middle ear).

Screening the sensory portion of the facial nerve. Screen the sense of taste on the anterior tongue.

- Place the patient in a seated position.

- Occlude vision.

- Present sweet, salty, and sour solutions to the lateral portions of the anterior tongue.

- Apply the taste substances first to the uninvolved side of the tongue, then to the involved side.

- Present each taste substance one at a time. Use a separate cotton applicator for each taste substance.

- Ask the patient to indicate whether he or she can taste the substance and identify whether it is sweet, sour, or salty.

☞ Impairment is indicated if the patient is unable to detect taste on one or both sides of the anterior tongue.

Note: The tip of the tongue can detect all tastes but detects primarily sweet tastes.

Screening the motor portion of the facial nerve. Screen the strength and symmetry of the facial muscles.

- Sit facing the patient, who also is seated.

- Ask the patient to elevate his or her eyebrows and forehead. The ability to wrinkle the forehead is used to distinguish an *upper motor neuron (UMN)* lesion from a *lower motor neuron (LMN)* lesion. With an UMN lesion, the muscles of the forehead will remain intact, even while the lower facial regions are impaired. This commonly occurs in cerebrovascular accidents. Bell's palsy is an LMN disorder in which the facial nerve has been lesioned. Both the forehead and lower facial muscles are paralyzed.

- Ask the patient to smile, frown, and pucker the lips.

- Check for asymmetry on the right and left sides of the face.

- Ask the patient to blow his or her cheeks up with air. Gently push on the patient's cheeks while asking him or her to resist (maintain air inside of the cheeks; see Figure 7.3).

- Check for asymmetry of facial muscle strength.

☞ Impairment is indicated if the patient has difficulty elevating the eyebrows and forehead on the involved side. Impairment also is indicated by decreased muscle strength and asymmetry on the involved side.

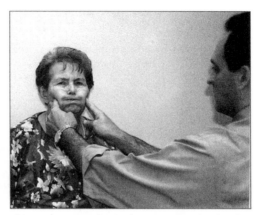

Figure 7.3 Gently Push on Patient's Cheeks: Screening Motor Portion of Facial Nerve

Note: Because the facial nerve travels through the internal auditory meatus and later gives off branches to the stapedius muscle, patients with Bell's palsy may complain of *ipsilateral hyperacusis* (increased sensitivity to sound).

CRANIAL NERVE 8: VESTIBULOCOCHLEAR NERVE

The vestibulocochlear nerve has an auditory branch and a vestibular branch. The *auditory branch* is responsible for audition (hearing). The auditory branch transmits sensory impulses that result from the vibrations of the fluid in the cochlea. The *vestibular branch* plays an important role in balance/equilibrium and the position of the head in space. The vestibular branch receives sensory stimulation from the semicircular canals in the inner ear.

Screening the auditory branch. To screen the auditory branch, an audiologist must first distinguish whether the patient has one of two types of hearing impairment:

1. *Sensorineural* impairment involves the inner ear, vestibulocochlear nerve, and the brain.

2. *Conductive* impairment involves the outer and middle ear structures.

To determine whether a hearing impairment is sensorineural or conductive, use the following procedures:

- Strike a tuning fork and place it against the middle of the patient's forehead (Weber test; see Figure 4.1).

☞ Sound is heard through bone conduction and should be heard equally in both ears. If sound is heard louder in one ear, this may indicate a conductive hearing loss.

- Strike a tuning fork and place it 1 inch from the auditory canal (Rinne test; see Figure 4.2).

- Place the vibrating tuning fork against the mastoid bone.

- Alternate placement until no sound is heard by the patient.

☞ Sound conducted by air is heard longer than sound conducted by bone. If the patient hears the sound longer when the tuning fork is placed on the mastoid bone, sensorineural hearing loss is indicated.

Screening the vestibular branch—presence of nystagmus.

- Place the patient in a seated position.

- Ask the patient to maintain his or her head in a fixed position while visually tracking a moving stimulus (a pen cap) held at a distance of 15 inches from the patient's face.

- Move the pen cap in an *H* and *X* pattern.

- Check for nystagmus both within and at the end ranges of the visual fields.

☞ Impairment is indicated if nystagmus is detected in the temporal or nasal visual fields.

Screening the vestibular branch—balance and the presence of protective responses (Rhomberg test).

- Ask the patient to assume a standing position with his or her eyes open.

- Check for increased sway and loss of balance.

- Gently displace the patient's balance.

- Check for the appropriate presence of protective responses (see Figure 4.3).

- Repeat the procedure with the patient's eyes closed.

☞ Impairment is indicated if the protective responses are absent or sluggish.

Screening the vestibular branch—presence of extensor tone in the lower extremities.

- Attempt to move the patient's lower extremity through its full range of motion.

- Screen one lower extremity at a time. Note whether extensor tone is present at the hip, knee, and ankle joints.

- Perform quick stretches of the hip, knee, and ankle to elicit the presence of extensor tone (see Section 5, "Motor Screening Procedures").

☞ Increased tone in the lower extremities may indicate impairment in the vestibular branch of the vestibulocochlear nerve.

CRANIAL NERVES 9 AND 10: GLOSSOPHARYNGEAL AND VAGUS

The glossopharyngeal nerve has both sensory and motor functions. The sensory portion of the glossopharyngeal nerve is responsible for taste on the posterior aspect of the tongue. The motor portion is responsible for swallowing and speaking. The vagus nerve has visceral and skeletal muscle branches. The visceral branches carry parasympathetic information to and from the heart, pulmonary system, esophagus, and gastrointestinal tract. The skeletal muscle branches carry motor information to the muscles of the larynx, pharynx, and upper esophagus. These muscles are responsible for swallowing and speaking. Both the glossopharyngeal and vagus nerves are responsible for the gag reflex (touching of the pharynx elicits contraction of the pharyngeal muscles) and swallowing reflex (food touching the pharynx elicits movement of the soft palate and contraction of the pharyngeal muscles). The glossopharyngeal and vagus nerves are screened together.

Screening the sensory portion of the glossopharyngeal nerve. Screen the sense of taste on the posterior aspect of the tongue.

- Place the patient in a seated position.

- Occlude vision.

- Present a bitter solution (e.g., rind of a lemon) to the posterior portion of the tongue (where bitter tastes can be detected) with a cotton swab.

- Apply the taste substance first to the uninvolved side of the tongue, then to the involved side.

- Ask the patient to indicate whether he or she can taste the substance and identify what the taste substance is. Give the patient a choice of three answers if the patient has word-finding difficulties.

- Ask the patient to identify whether the taste substance is sweet, sour, salty, or bitter.

☞ Impairment is indicated if the patient is unable to detect taste on one or both halves of the posterior region of the tongue.

Screening the motor portions of the glossopharyngeal and vagus nerves.

- Attempt to elicit the gag reflex by swiping a tongue depressor or cotton swab at the back of the patient's throat.

- Observe the patient's ability to swallow different consistencies of food (dysphagia).

- Present different consistencies of food one at a time (solid foods, pureed foods, thick liquids, thin liquids).

- Ask the patient to consume each presented food type. Check for food aspiration, coughing, throat clearing, or a wet vocal quality (indicating that the food is pocketing in the larynx). *Note:* If aspiration precautions have been indicated, modify the above procedure accordingly.

- Observe the patient's ability to speak clearly without slurring words (dysarthria).

- Check for decreased phonal volume or hoarseness of voice (dysphonia). Because the visceral branches of the vagus nerve carry information to and from the pulmonary system, the patient should also be checked for breathing difficulties (dyspnea).

☛ Impairment is indicated if the patient has difficulty swallowing and aspirates food. Impairment also is indicated if the patient presents with dysarthria, dysphonia, and/or dyspnea.

CRANIAL NERVE 11: ACCESSORY NERVE

The accessory nerve has two nerve roots: a *cranial nerve root* and a *spinal nerve root*. The cranial nerve root controls the larynx during swallowing. The spinal nerve root innervates the sternocleidomastoid and upper trapezius muscles. Innervation of the sternocleidomastoid muscle allows for head rotation to the contralateral side and head flexion–extension. Innervation of the upper trapezius muscle allows for shoulder elevation and shoulder flexion above 90°.

Screening the cranial nerve root.

- Place the patient in a seated position.

- Place index and middle fingers over the patient's Adam's apple (laryngeal muscles).

- Instruct the patient to swallow. Check for the normal rise and fall of the larynx.

- Check for the presence of dysphagia by screening the patient's ability to handle different food consistencies (as described in the section on glossopharyngeal and vagus nerves).

☛ Impairment is indicated if the larynx does not rise and fall normally as the patient swallows. Impairment also is indicated by the presence of dysphagia.

Screening the spinal nerve root—sternocleidomastoid muscle.

- Instruct the patient to flex his or her head laterally and forward and to rotate the head to the opposite side. Ask the patient to resist the your attempts to prevent the above movements.

- Screen the uninvolved side first, followed by the involved side. Compare both sides of the body to observe the patient's symmetry of movement and strength.

- Check for atrophy of the involved sternocleidomastoid muscle.

☞ Impairment is indicated by weakness or atrophy of the sternocleidomastoid muscle.

Screening the spinal nerve root—upper trapezius muscle.

- Instruct the patient to shrug both shoulders toward his or her ears. Ask the patient to resist your attempts to depress his or her elevated shoulders.

- Screen the uninvolved side first, followed by the involved side. Check for symmetry in shoulder movement and strength.

- Check for atrophy of the involved upper trapezius muscle.

☞ Impairment is indicated by weakness or atrophy of the upper trapezius muscle.

CRANIAL NERVE 12: HYPOGLOSSAL NERVE

The hypoglossal nerve is responsible for tongue movement.

- Ask the patient to protrude his or her tongue. Note whether the tongue deviates to the side of the lesion (see Figure 7.4) and whether unilateral or bilateral atrophy of the tongue muscles exist.

- Check for tongue tremors or involuntary tongue movements.

- Instruct the patient to move his or her tongue from side to side.

- Check for asymmetry in movement and weakness.

Figure 7.4 Tongue Deviation: Screening Hypoglossal Nerve

- Instruct the patient to push his or her tongue against one cheek and then the other.

- Ask the patient to resist your attempts to depress the cheek as he or she pushes the cheek outward. Note whether asymmetry in tongue strength exists.

☞ Impairment is indicated by tongue deviation toward the side of the lesion, unilateral or bilateral atrophy of the tongue muscles, tongue tremors, and/or unilateral or bilateral weakness of the tongue muscles.

RED FLAGS
SIGNS AND SYMPTOMS OF CRANIAL NERVE DAMAGE

A red flag is an indicator or symptom of dysfunction. If a patient is observed displaying any of the following red flags, it is likely that the patient possesses cranial nerve damage.

- **OLFACTORY NERVE DAMAGE** may result in anosmia, or loss of smell.

- **OPTIC NERVE DAMAGE** may result in unilateral blindness.

- **OCULOMOTOR NERVE DAMAGE** may result in lateral strabismus (or exotropia), ptosis of the eyelid, loss of the pupillary reflex, or impaired visual convergence and consensuality.

- **TROCHLEAR NERVE DAMAGE** may result in vertical, medial strabismus that prevents the patient from looking down.

- **ABDUCENS NERVE DAMAGE** may result in medial strabismus.

- **EXTRAOCULOMOTOR NERVE DAMAGE** (cranial nerves 3, 4, and 6) may result in diplopia or double vision (often the result of some form of strabismus) or nystagmus.

- **TRIGEMINAL NERVE DAMAGE** may result in loss of the corneal reflex; weakness in chewing (weakness in the muscles of mastication); deviation of jaw to affected side; or loss of sensation to the head, face, and inner oral cavity. When one half of the face loses sensation, trigeminal neuralgia may be indicated.

- **FACIAL NERVE DAMAGE** may result in decreased or lost taste sensation on the anterior tongue, decreased or lost corneal reflex (blink reflex), or paralysis or weakness of the facial muscles (on the side of the lesion). Drooping or paralysis of the ipsilateral side of the face often indicates Bell's palsy.

- **GLOSSOPHARYNGEAL NERVE DAMAGE** may result in loss of taste sensation on the posterior aspect of the tongue (loss of bitter taste modality), loss of the gag and swallowing reflexes, or dysphagia (difficulty swallowing, leading to choking or food aspiration).

- **VAGUS NERVE DAMAGE** may result in dyspnea (difficulty breathing) if the visceral branch of the vagus nerve is damaged. Dysphonia (hoarse voice), dysphagia (difficulty swallowing), or dysarthria (difficulty enunciating words, slurring words) may result from damage to the skeletal branch of the vagus nerve.

- **ACCESSORY NERVE DAMAGE** may result in dysphagia (secondary to laryngeal elevation) if the cranial nerve root of the accessory nerve is damaged. If the spinal nerve root of the accessory nerve is damaged, weakness rotating the head to the contralateral side, flexing the head laterally and forward, extending the head, elevating the shoulder (shrugging the shoulder) on the involved side, or flexing the arm above 90° (on the involved side) may result.

- **HYPOGLOSSAL NERVE DAMAGE** may result in dysarthria (secondary to impaired tongue musculature, causing an inability to produce tongue movements for sound and word formation), ipsilateral deviation of the tongue, dysphagia (because the tongue muscles are needed to manipulate food into a bolus in the mouth and propel the bolus to the pharynx), ipsilateral atrophy of the tongue, or ipsilateral paralysis of the tongue.

AVAILABLE IN-DEPTH ASSESSMENTS

If cranial nerve screening detects neurologic impairment, the following in-depth assessments are available:

Activities-Specific Balance Confidence Scale (ABC)
L. E. Powell, A. M. Myers
Adults with balance deficits
Powell, L. E., & Myers, A. M. (1995). The Activities-Specific Balance Confidence (ABC) Scale. *Journal of Gerontology, 50A*(1), M28–M34.

Bedside Evaluation of Dysphagia (BED)
E. Hardy
Adults with eating and swallowing deficits
Imaginart International, Inc.
307 Arizona St.
Bisbee, AZ 85603
(520) 432-5741

Behavioural Inattention Test (BIT)
B. Wilson, J. Cockburn, P. Halligan
Adults with visual neglect secondary to neurologic impairment
National Rehabilitation Services
117 North Elm St.
PO Box 1247
Gaylord, MI 49735
(517) 732-3866

Berg Balance Scale (Berg)
K. O. Berg, S. L. Wood-Dauphinee, J. I. Williams, B. Maki
Adults with balance deficits
Berg, K. O., Wood-Dauphinee, S. L., Williams, J. I., & Maki, B. (1992). Measuring balance in the elderly: Validation of an instrument. *Canadian Journal of Public Health, 83*, S7–S11.

Brain Injury Visual Assessment Battery for Adults (BIVABA)
M. Warren
Adults with visual deficits secondary to neurologic impairment
Precision Vision
745 North Harvard Ave.
Villa Park, IL 60181

Chessington OT Neurological Assessment Battery (COTNAB)
R. Tyerman, A. Tyerman, P. Howard, C. Hadfield
Adults with neurologic impairment
North Coast Medical
187 Stauffer Blvd.
San Jose, CA 95125-1042

Clinical Test of Sensory Interaction in Balance (CTSIB)
M. Watson
Adults with balance deficits
Watson, M. (1992). Clinical Test of Sensory Interaction in Balance. *Physiotherapy Theory and Practice, 8*, 176–178.

Comprehensive Test of Visual Functioning (CTVF)
S. L. Larson, E. Buethe, G. J. Vitali
Children, adolescents, and adults with neurologic impairment
Slosson Educational Publications, Inc.
PO Box 280
East Aurora, NY 14052-0280
(800) 828-4800

Dizziness Handicap Inventory (DHI)
G. P. Jacobson, C. W. Newman
Adults with balance deficits secondary to vestibular impairment
Jacobson, G. P., & Newman C. W. (1990). The development of the Dizziness Handicap Inventory. *Archives of Otolaryngology, Head, and Neck Surgery, 116*, 424–427.

Dysphagia Evaluation Protocol
W. Avery-Smith, A. Brod Rosen, D. M. Dellarosa
Adults with eating and swallowing disorders
Therapy Skill Builders
Psychological Corporation
555 Academic Ct.
San Antonio, TX 78204-2498
(800) 211-8378, (800) 228-0752

Elemental Driving Simulator (EDS) and Driving Assessment System (DAS)
R. Giatnutsos, A. Campbell, A. Beattie, F. Mandriota
Adults with neurologic impairment
Life Sciences Associates
One Fenimore Rd.
Bayport, NY 11795-2115
(516) 472-2111

Erhardt Developmental Vision Assessment (EDVA)
R. Erhardt
Individuals suspected of visual–motor deficits
Therapy Skill Builders
Psychological Corporation
PO Box 998
Odessa, FL 33556-9908
(800) 228-0752, (800) 232-1223

Evaluation of Oral Function in Feeding
M. Stratton
Children and adults with eating and swallowing disorders
Stratton, M. (1981). Behavioral assessment scale of oral functions in feeding. *American Journal of Occupational Therapy, 35,* 719–721.

Falls Efficacy Scale (FES)
M. E. Tinnetti, D. Richman, L. Powell
Elderly persons with balance deficits
Tinnetti, M. E., Richman, D., & Powell, L. (1990). Falls efficacy as a measure of fear of falling. *Journal of Gerontology, 45,* 239–243.

Functional Reach Test (FRT)
P. W. Duncan, D. K. Weiner, J. Chandler, S. A. Studenski
Adults with balance deficits secondary to neurologic damage
Duncan, P. W., Weiner, D. K., Chandler, J., & Studenski, S. A. (1990). Functional reach: A new clinical measure of balance. *Journal of Gerontology, 45,* 192–197.

Morse Fall Scale (MFS)
J. M. Morse
Adults with balance deficits
Janice M. Morse

Pennsylvania State University
School of Nursing
201 Health and Human Development East
University Park, PA 16802-6508

Physical Performance Test (PPT)
D. B. Reuben, A. L. Siu
Elderly persons with balance deficits
Reuben, D. B., & Siu, A. L. (1990). An objective measure of physical function of elderly outpatients: The Physical Performance Test. *Journal of the American Geriatric Society, 38,* 1105–1112.

Revised Sheridan Gardiner Test of Visual Acuity
M. D. Sheridan, P. Gardiner
Age 5 and over
Keeler Instruments, Inc.
456 Parkway
Broomall, PA 19008
(610) 353-4350

Structured Observational Test of Function (SOTOF)
A. Laver, G. E. Powell
Elderly persons with neurologic impairment or dementia
NFER-NELSON Publishing Company
Darville House
2 Oxford Rd. East
Windsor, Berkshire SL4 1DF England
0753-858961

Timed Get Up and Go (TGUG)
D. Podsiadlo, S. Richardson
Adults with balance deficits
Podsiadlo D., & Richardson, S. (1991). The Timed Get Up and Go: A test of basic functional mobility for frail elderly persons. *Journal of the American Geriatric Society, 39,* 142–148.

Tinnetti Balance Test of the Performance-Oriented Assessment of Mobility Problems (Tinnetti)
M. E. Tinnetti
Adults with balance deficits

Tinnetti, M. E. (1986). Performance-oriented assessment of mobility in elderly patients. *Journal of the American Geriatric Society, 34,* 119–126.

Visual Attention Test
D. A. Lister
Adults with visual attention impairment secondary to stroke
Lister, D. A. (1984). Apparatus for assessing vision after stroke. *Canadian Journal of Occupational Therapy, 51,* 237–240.

For further information about the these assessments, refer to the following resources:

Asher, I. E. (1996). *Occupational therapy assessment tools: An annotated index* (2nd ed.). Bethesda, MD: American Occupational Therapy Association.

Impara, J. C., Plake, B. S., & Murphy L. L. (Eds.). (1998). *The thirteenth mental measurements yearbook.* Lincoln: University of Nebraska Press.

Murphy, L. L., Impara, J. C., & Plake, B. S. (Eds.). (1999). *Tests in print: An index to tests, test reviews, and the literature on specific tests* (Vols. 1 & 2). Lincoln: University of Nebraska Press.

Plake, B. S., Impara, J. C., & Murphy, L. L. (Eds.). (1999). *The supplement to* The Thirteenth Mental Measurements Yearbook. Lincoln: University of Nebraska Press.

Because a dysphagia evaluation can fail to detect a swallowing problem, a referral for a *videofluoroscopy evaluation* should always be made when cranial nerve damage has been indicated by screenings or neuroimaging scans.

Referral to an *ophthalmologist* should be made when screening has detected significant visual impairment.

Referral to an *audiologist* for an in-depth hearing exam should be made when screening has detected auditory impairment.

Referral to a *neurologist* for an in-depth neurologic workup—using imaging technology—is indicated when screening has detected possible neuropathology.

CRANIAL NERVE SCREENING

Patient Name _____   Date _____

CRANIAL NERVE 1: OLFACTORY NERVE

☐ No pathology detected   ☐ Possible pathology present:

        ☐ unilateral loss of smell   ☐ bilateral loss of smell

---

CRANIAL NERVE 2: OPTIC NERVE

Results of Snellen Eye Chart

    Right eye _____   Left eye _____   Binocular vision _____

Visual field screening (confrontation)

    Peripheral vertical vision _____ (45°)   Horizontal/temporal vision _____ (85°)

CRANIAL NERVE 3: OCULOMOTOR NERVE

Pupillary constriction

    Right eye: ☐ normal constriction   ☐ possible pathology present

    Left eye: ☐ normal constriction   ☐ possible pathology present

Consensuality

    Right eye: ☐ normal constriction   ☐ possible pathology present

    Left eye: ☐ normal constriction   ☐ possible pathology present

    Binocular: Both eyes constrict together

        ☐ to same degree   ☐ possible pathology present

Convergence

    Occurred at _____ (3–4") distance from patient's nose

    Did pupils constrict as the eyes converged?   ☐ Yes   ☐ No

---

CRANIAL NERVE 3, 4, & 6: OCULOMOTOR, TROCHLEAR, & ABDUCENS

Scanning

    Right eye: ☐ normal scanning   ☐ possible pathology present

    Left eye: ☐ normal scanning   ☐ possible pathology present

    Binocular: ☐ normal scanning   ☐ possible pathology present

Is a strabismus detected in either eye?   ☐ Yes ☐ No ☐ Right eye ☐ Left eye

Does the patient report diplopia (double vision)? ☐ Yes ☐ No ☐ Right eye ☐ Left eye

Is a pathological nystagmus detected? ☐ Yes ☐ No ☐ Right eye ☐ Left eye ☐ Both

        ☐ Horizontal   ☐ Vertical   ☐ Rotary

CRANIAL NERVE 5: TRIGEMINAL

Right-side sensation	Sharp	Dull	Temp
Face	☐ Normal	☐ Normal	☐ Normal
	☐ Pathology	☐ Pathology	☐ Pathology
Head	☐ Normal	☐ Normal	☐ Normal
	☐ Pathology	☐ Pathology	☐ Pathology
Inner oral cavity	☐ Normal	☐ Normal	☐ Normal
	☐ Pathology	☐ Pathology	☐ Pathology

Left-side sensation	Sharp	Dull	Temp
Face	☐ Normal	☐ Normal	☐ Normal
	☐ Pathology	☐ Pathology	☐ Pathology
Head	☐ Normal	☐ Normal	☐ Normal
	☐ Pathology	☐ Pathology	☐ Pathology
Inner oral cavity	☐ Normal	☐ Normal	☐ Normal
	☐ Pathology	☐ Pathology	☐ Pathology

Corneal reflex

    Right eye: ☐ Normal   ☐ Possible pathology present

    Left Eye: ☐ Normal   ☐ Possible pathology present

CRANIAL NERVE 7: FACIAL

Taste sense on anterior tongue

Right side: ☐ Normal   ☐ Possible pathology present

Left side: ☐ Normal   ☐ Possible pathology present

Strength and symmetry of facial muscles

Right side: ☐ Normal   ☐ Possible pathology present

Left side: ☐ Normal   ☐ Possible pathology present

CRANIAL NERVE 8: VESTIBULOCOCHLEAR

Hearing

Right side: ☐ Normal   ☐ Possible pathology present

Left side: ☐ Normal   ☐ Possible pathology present

Is a pathological nystagmus detected?   ☐ Right eye   ☐ Left eye   ☐ Both eyes

Presence of protective responses   ☐ Normal   ☐ Possible pathology present

Is abnormal extensor tone detected in the lower extremities?   ☐ Yes   ☐ No   ☐ Right   ☐ Left

CRANIAL NERVE 9 & 10: GLOSSOPHARYNGEAL & VAGUS

Taste sense on posterior aspect of tongue

Right side: ☐ Normal   ☐ Possible pathology present

Left side: ☐ Normal   ☐ Possible pathology present

Presence of gag reflex   ☐ Normal   ☐ Possible pathology present

Possible dysphagia detected?   ☐ Yes   ☐ No

Presence of dysarthria detected?   ☐ Yes   ☐ No

Presence of dysphonia (hoarseness) detected?   ☐ Yes   ☐ No

Presence of dyspnea (breathing difficulties) detected?   ☐ Yes   ☐ No

CRANIAL NERVE 11: ACCESSORY

Possible dysphagia present?   ☐ Yes   ☐ No

Sternocleidomastoid muscle function:

Right side:

Flex head laterally   ☐ Normal   ☐ Pathology

Flex head forward   ☐ Normal   ☐ Pathology

Rotate head to opposite side   ☐ Normal   ☐ Pathology

Left side:

Flex head laterally   ☐ Normal   ☐ Pathology

Flex head forward   ☐ Normal   ☐ Pathology

Rotate head to opposite side   ☐ Normal   ☐ Pathology

Upper trapezius muscle function:

Shoulder elevation

Right side: ☐ Normal   ☐ Pathology

Left side: ☐ Normal   ☐ Pathology

CRANIAL NERVE 12: HYPOGLOSSAL

Tongue movement

Right side: ☐ Normal   ☐ Pathology

Left side: ☐ Normal   ☐ Pathology

Tongue atrophy

Right side: ☐ Normal   ☐ Pathology

Left side: ☐ Normal   ☐ Pathology

Neuropathy and Peripheral Nerve Functioning
SCREENING

A neuropathy involves some form of impairment occurring within the peripheral nervous system. The peripheral nervous system includes all neural structures that lie outside the pia mater and the brain and spinal cord. The optic and olfactory nerves are not included because they are special extensions of the brain that lie within the oligodendroglia and are supported by astrocytes. In contrast, peripheral nervous system structures are wrapped with myelin sheaths that are enclosed by Schwann cells.

Thus, the peripheral nervous system includes the motor and sensory spinal nerves, the dorsal root and rootlets, the dorsal root ganglia, the ventral horn, and the ventral root and rootlets. Both the cranial nerves (except for cranial nerves 1 and 2) and the cauda equina also are considered to be part of the peripheral nervous system.

The term *neuropathy* refers to impairment of any of the peripheral nervous system structures. Impairment may take the form of the following:

- A disease process (e.g., anterior horn cell syndrome, diabetes mellitus, Guillain-Barré syndrome)

- A peripheral nerve injury—either severance or compression (e.g., radial nerve damage causing wrist drop, median nerve damage carpal tunnel)

- An entrapment syndrome involving the plexi and/or spinal nerve roots (e.g., suprascapular nerve entrapment, sciatica)

- A vascular condition (ischemia) that causes denervation (e.g., thoracic outlet syndrome).

There are five major categories of peripheral neuropathy:

- *Polyneuropathy* involves bilateral, symmetric damage to the peripheral nerves. Usually the lower extremities are more commonly involved than are the upper extremities. The distal segments of an extremity are involved first, before the proximal segments. Signs and symptoms commonly start in the feet and move proximally to the knees. Often, symptoms are present at the knee level before they are found in the hands. The typical pattern of presentation is stocking-and-glove neuropathy in which the feet and knees are affected first, followed by the hands. One etiology causing stocking-and-glove neuropathy is diabetes mellitus.

- *Mononeuropathy* involves damage to a single nerve, for example, radial nerve damage causing wrist drop and peroneal nerve damage causing foot drop.

- *Mononeuropathy multiplex* involves damage of multiple single peripheral nerves in an asymmetric random pattern. Common causes include diabetes mellitus and polyarteritis.

- *Plexopathy* involves damage to the nerves in one of the plexi—brachial, lumbar, or sacral plexus. Idiopathic brachial neuritis and traumatic injury to the brachial plexus are common causes.

- *Radiculopathy* involves spinal nerve root damage to the dorsal or ventral roots. Herniated vertebral discs and vertebral bone disease are common causes.

Peripheral neuropathies can be further broken down into two subcategories:

- *Axonal neuropathies* involve damage to the nerve's cell body or axon, the structure through which neuronal signals travel. One example is anterior horn cell syndrome, in which damage to the cell bodies of the ventral horn produce motor weakness with atrophy and fasciculations.

- *Demyelinating neuropathies* involve damage to the myelin sheath surrounding the axon, the structure that controls the speed of nerve signal conduction. One example is Guillain-Barré syndrome, or acute inflammatory polyneuropathy, in which demyelination causes bilateral, symmetric weakness that begins in the distal lower extremities and progresses proximally.

Peripheral nerve damage can be subdivided into three classes:

- *Class 1 neuropraxia.* A rapidly reversible blockage resulting from transient ischemia with no anatomical changes to the nerve. Symptoms include decreased strength, absence of deep tendon reflexes, and hypo- and hyperesthesia or sensation. Recovery occurs within 3 months.

- *Class 2 axonotmesis.* The nerve axon is damaged secondary to closed-crush or percussion injuries. Schwann cells remain intact. Wallerian degeneration occurs distal to the lesion site. Symptoms include loss of sensation, motor, and sympathetic nerve function distal to the lesion site; muscular atrophy; areflexia; and possible trophic changes to the skin and nails. Regeneration is high because of the integrity of the Schwann cells but is slow, taking several months to 1 year (8 mm per day).

- *Class 3 neurotmesis* is complete severance of the nerve secondary to stab wounds or other high-velocity projectiles. Wallerian degeneration occurs distal to the lesion site. Symptoms include loss of sensation, motor, and sympathetic nerve function distal to the lesion site; muscular atrophy; areflexia; and possible trophic changes to the skin and nails. Regeneration may occur, but the prognosis for complete recovery is poor.

FUNCTIONAL IMPLICATIONS OF NEUROPATHY AND PERIPHERAL NERVE DAMAGE

SENSORY DISTURBANCES

Peripheral neuropathies and nerve disorders typically produce both negative and positive phenomena. Negative phenomena involve an absence or decrease in the perception of sensation—referred to as *hypoesthesia*. Positive phenomena involve an increase in the perception of sensation—referred to as *hyperesthesia*. *Paresthesia* is the occurrence of unusual feelings such as pins and needles. *Dysesthesia* is the occurrence of unpleasant sensations such as burning. *Causalgia* is an intense burning pain accompanied by trophic skin changes (on the region of denervation, the skin becomes smooth and glossy, the nails become thick, and hair thins or falls out).

Discriminative touch, proprioception, and vibration are associated with large-diameter fibers that are more vulnerable to compression injuries than are small-diameter fibers. If discriminative touch, proprioception, and vibration are impaired, the injury is thought to involve large-diameter fibers. Small-diameter fibers are associated with the sensation of temperature and pain and are less vulnerable to compression injuries. If temperature and pain are impaired, the injury is thought to involve small-diameter fibers.

Any absence or decrease in the perception of sensory stimuli diminishes one's safety. Individuals with decreased or absent sensation secondary to neuropathy are at risk for burns that may occur during meal preparation activities or bathing. Patients with lower-extremity neuropathy (involving decreased sensation) are at a greater risk for falls during all activities of daily living. The use of knives or any sharp utensils (the blades of a food processor) also can become safety hazards for patients with decreased sensation secondary to neuropathy. Similarly, driving becomes a precarious activity if one cannot readily feel through feet or hands. If proprioceptive feedback is lost as a result of neuropathy, the exact pressure required to manipulate the accelerator, brake pedal, and steering wheel of a car becomes uncertain.

Patients also may demonstrate neglect of their insensitive area. Patients may fail to perform proper hygiene of their insensitive area or may inadvertently cause repeated trauma to that area through poor safety habits. Because patients have a tendency to hold their affected limb in a guarded or dependent position, they may cause increased swelling, or edema, to occur. Conversely, the experience of hyperesthesia may be so painful that it disrupts the patient's ability to function in daily activities. The feel of water pellets from a shower or the sensation of a wet washcloth may be intolerable to a patient with hyperesthesia, thus making bathing difficult.

The pain of hyperesthesia may be so extreme that it interferes with the patient's ability to attend work or school. Many peripheral neuropathies that are experienced as painful prevent the patient from participating in the most basic activities. Even walking across a room to get a drink of water or rising from a chair to go to the bathroom become strenuous, unpleasant tasks. Consequently, patients experience further muscular atrophy from disuse.

MOTOR DISTURBANCES

The specific pattern of weakness and paralysis seen with neuropathy and peripheral nerve disorder is directly related to the site and severity of the lesion or disease process. Sites particularly vulnerable to injury and entrapment syndromes include the *peroneal nerve* (near the proximal fibula), the *median nerve* (at the carpal tunnel), and the *radial and ulnar nerves* (as they travel past the radial groove and medial epicondyle of the humerus).

Muscular weakness occurs as a result of partial denervation. Flaccidity results from complete denervation. When muscular weakness or flaccidity occur, atrophy appears shortly thereafter. Patients commonly report rapid fatigue and decreased endurance—both contributing to the patient's diminished ability to perform ADL. Muscular cramping commonly occurs with fatigue. Along with flaccidity, deep tendon reflexes become hyporeflexive. As a result of fatigue and decreased endurance, patients may require a lengthier than normal amount of time to complete basic ADL—bathing, grooming, dressing, meal preparation. Such patients must often learn energy conservation strategies to prevent or decrease fatigue during ADL.

Patients with neuropathy secondary to a disease process (e.g., Guillain-Barré syndrome, diabetes mellitus, amyotrophic lateral sclerosis) may find bathing, grooming, dressing, and meal preparation to be impossible without assistance or adaptive equipment. Neuropathies and peripheral nerve disorders in which proprioceptive feedback is lost will contribute to balance problems. Standing in the shower, leaning over to put on socks and shoes, and reaching down below the trunk or high above the head to retrieve a food item from a kitchen cabinet or the refrigerator are ADL that place the patient at risk for falls and injury. Protective and equilibrium responses will be diminished or absent, further placing the patient at risk for falls and injury. Neuropathies in which proprioceptive feedback is lost also will impair one's ability to drive; to pick up and carry groceries or laundry; and to lower and raise oneself from a chair, toilet, and bed.

AUTONOMIC DYSFUNCTION

Autonomic dysfunction may result in vasomotor nerve disturbances. Vasomotor nerves innervate the motor activity of blood vessel walls. Such nerves control blood vessel vasodilation and constriction. Vasoconstriction can lead to ischemia of a specific body region. One example of a neuropathy resulting from ischemia is *thoracic outlet syndrome* in which the presence of a cervical rib (or some other narrowing of the thoracic outlet) impinges blood supply of the median nerve. Autonomic dysfunction can also lead to alterations in sweating—the patient experiences either an abnormal increase or decrease in sweating of the denervated region. Trophic (or nutritional) changes often accompany autonomic nerve dysfunction. The patient's denervated skin region may become smooth and glossy. Nails in a denervated body region often thicken abnormally. Similarly, hair over a denervated skin region will thin or fall out. All of these symptoms of autonomic nerve dysfunction can be physically uncomfortable and unsightly for the patient, causing him or her to feel embarrassed in and to avoid social settings.

SOFT TISSUE CHANGES

Connective tissue changes can occur secondary to disuse after paralysis and/or joint immobilization. Often, tendons thicken and fibrotic adhesions form in the periarticular regions. Such soft tissue changes can decrease a patient's joint range of motion (ROM) and ultimately limit a patient's functional recovery after reinnervation. Limitations in joint ROM can adversely affect all ADL in which full joint ROM is critical (e.g., bathing, dressing, driving).

Joint weakness and instability may develop when the muscles surrounding a joint become paralyzed. Over time, edema and disuse further weaken a joint capsule and ligaments, eventually predisposing the involved joints to hypermobility. Hypermobility and joint instability make a joint vulnerable to subluxation and further damage articular surfaces. Subluxed joints and joints with deteriorated surfaces make movement painful and further limit functional use of the involved extremity.

NEUROPATHY SCREENING PROCEDURES

All screening for neuropathy should include (a) sensory screening of the involved regions, (b) ROM screening of the involved joints, (c) manual muscle testing (with comparison of the denervated muscles to the intact muscles), and (d) observation of how the patient's performance of functional activities has been affected by the presence of neuropathy. Additionally, the patient should receive electromyography (EMG) testing administered by a neurologist.

SENSORY SCREENING

An accurate sensory screening should not be tedious; rather, sensory screening should focus on the areas relevant to the diagnosis. Often, the patient can best indicate the region of sensory impairment. The area should then be examined in detail for discriminative touch, proprioception, vibration, and pain and temperature. Because large-diameter fibers are more vulnerable to injury than are small-diameter fibers, the impairment of specific types of sensory data offers an indication of the extent and type of the nerve damage. Discriminative touch, proprioception, and vibration are sensations that are transmitted along large-diameter fibers. Pain and temperature are sensations transmitted along small-diameter fibers.

Sensory examination should start with the most distal area of possible impairment and move proximally because nerve damage that follows a peripheral distribution affects the most distal body areas first and moves proximally. Often in polyneuropathy, the feet and lower legs are affected first, followed by the hands. The therapist should then test more proximal regions only if the patient does not perceive sensation distally.

If the nerve damage follows a true dermatomal pattern—rather than a peripheral nerve distribution—specific skin regions will be affected, depending on the specific peripheral nerve damaged, which often is the case in mononeuropathy multiplex. The therapist should examine the patient's ability to detect sensation in accordance with the dermatomal skin segments. Again, the patient often is best able to initially indicate the location of sensory impairment.

A tinel sign is a sensory screen used to determine the progression of nerve regeneration. The test involves tapping on the end of the regenerating nerve. A positive sign is indicated by pain or tingling in response to tapping over the distal end of the regenerating nerve. The patient cannot detect sensation in those areas still lacking innervation. The therapist also should closely follow any patient who develops hypersensitivity in response to nerve damage. The development of *reflex sympathetic dystrophy*—a disorder in which an upper extremity having nerve damage becomes immobile or frozen—is both highly painful and functionally limiting to the patient.

☛ The following indicate sensory impairment:

- Loss of discriminative touch, proprioception, and vibration in specific body regions indicates damage to large-diameter nerve fibers.

- Loss of pain and temperature in specific body regions indicates damage to small-diameter fibers. Because small-diameter nerve fibers are less vulnerable to injury than are large-diameter fibers, the loss of pain and temperature also indicates an injury of greater severity.

- Nerve damage that begins distally and moves proximally is referred to as damage that follows a true peripheral nerve distribution. Polyneuropathies resulting from denervated nerves follow a true peripheral nerve distribution.

- Nerve damage that affects distinct specific skin regions in accordance with dermatomal skin segments is said to follow a true dermatomal pattern.

- A positive tinel sign is indicated by pain or tingling in response to tapping over the distal end of a regenerating nerve.

- The presence of reflex sympathetic dystrophy is indicated by immobility and pain in an upper extremity having nerve damage. Reflex sympathetic dystrophy also is referred to as *frozen shoulder.*

- The presence of *paresthesia* (the occurrence of unusual feelings such as pins and needles) and *dysesthesia* (the occurrence of unpleasant sensations such as causalgia, or intense burning) is indicated along the denervated nerve.

SCREENING FOR AUTONOMIC DYSFUNCTION

The patient's denervated skin region should be examined for the presence of autonomic dysfunction over a denervated area, including alterations in sweating, presence of trophic (or nutritional) changes, smooth and glossy skin, abnormally thickened nails, and thinning or loss of hair.

SCREENING FOR SOFT TISSUE CHANGES

The patient's skin should be inspected for signs of neglect and breakdown. Subcutaneous layers should be palpated for the presence of edema, particularly in limbs held in a guarded, dependent position. Deep palpation should be performed to assess the condition of fascia, ligaments, and tendons.

☞ Soft tissue abnormalities secondary to nerve damage are indicated by the presence of

- Edema

- Adhesions and fascia thickening that limits joint ROM

- Joint instability and weakness resulting in joint hypermobility

- Atrophy or flaccidity of denervated muscles.

SCREENING FOR ROM DISTURBANCES

The therapist should assess the patient's active ROM and passive ROM. ROM screening should focus on involved joints. Comparison of the ROM of involved joints with uninvolved joints should be made if possible (i.e., if the patient's nerve damage does not involve symmetric polyneuropathy). Edema, tissue adhesions, and muscle denervation secondary to nerve damage will limit a patient's ROM. Conversely, joint weakness and instability may develop when the muscles surrounding a joint become paralyzed. Over time, edema and disuse further weaken a joint capsule and ligaments, eventually predisposing the involved joints to hypermobility. Hypermobility and joint instability make a joint vulnerable to subluxation and further damage articular surfaces. The therapist also should assess the patient's joint *end-feels*—the sensation of a joint when it is passively moved to the end of its range. The end-feel of joints with deteriorated surfaces can be felt as bone moving over bone rather than as slight tissue give felt when joint ligaments are moved to the end of their range.

☞ ROM disturbances are indicated by

- Limitations caused by tissue adhesions, edema, and/or muscle denervation secondary to nerve damage

- Joint weakness and instability caused by muscle denervation

- Hypermobility caused by edema and muscle atrophy

- Subluxation secondary to joint tissue damage

- Damaged articular surfaces secondary to joint capsule deterioration

- The presence of bony end-feels of involved joints.

SCREENING FOR MOTOR DYSFUNCTION

Manual muscle testing (MMT)—of involved muscles—should be performed to

- Determine whether nerve damage follows a myotomal pattern or peripheral nerve distribution

- Determine whether nerve damage is unifocal or multifocal

- Assess the completeness and severity of nerve damage.

The therapist should assess the patient's physical endurance of involved muscles because partially denervated muscles may demonstrate near-normal strength through MMT but fatigue quickly. Patients often complain of muscle cramping in partially denervated muscles. Deep tendon reflex testing (of involved limbs) should be performed because deep tendon reflexes will become hyporeflexive as a result of nerve damage.

☛ Motor dysfunction is indicated by

- Weakness and atrophy of involved muscles

- Reports of muscle cramping

- Hyporeflexia.

ELECTROMYOGRAPHY

An EMG examination should be administered by a neurologist. EMG can demonstrate the presence or absence of normal innervation to a muscle and consists of two components: (a) an evaluation of nerve conduction and (b) an evaluation of the electrical activity in muscles.

PERIPHERAL NERVE SCREENING PROCEDURES

The following describe the screening procedures for the most commonly experienced peripheral nerve disorders of the upper and lower extremities.

COMMON PERIPHERAL NERVE SYNDROMES OF THE UPPER EXTREMITY

1. *Upper brachial plexus injury—Erb-Duchene syndrome and waiter's tip position.* This disorder arises from the C5 and C6 spinal nerve roots, which innervate the muscles responsible for scapular, shoulder, elbow, and forearm movements.

The brachial plexus arises from C5, C6, C7, C8, and T1 peripheral nerve roots. Erb-Duchene syndrome is the most common injury to the upper brachial plexus and occurs as a result of damage to C5 and C6 peripheral nerve roots. Common causes of injury are traction injuries (occurring at birth) and compression injuries. Upper plexus injuries can cause paralysis of the deltoid, biceps, brachialis, and brachioradialis muscles. Sometimes the supraspinatus, infraspinatus, and subscapularis muscles also are affected. If the peripheral nerve roots are avulsed (torn away) from the spinal cord, the rhomboids, serratus anterior, levator scapula, and the scalene muscles likely will be affected.

With an upper brachial plexus injury, the affected arm is held limply at the patient's side in internal rotation and adduction of the shoulder. The elbow is extended. The forearm is pronated.

This is referred to as the *waiter's tip position* (see Figure 8.1). Biceps and brachioradialis reflexes are lost, and sensation in the deltoid and radial surfaces of the forearm and hand is lost.

Biceps tendon reflex screening:

- Place the patient's arm in 90° of elbow flexion across your forearm.

- Place hand under the medial aspect of the patient's elbow.

- Place thumb on the biceps tendon in the cubital fossa of the patient's elbow. To locate the tendon, ask the patient to tense the biceps muscle as your thumb continues to rest in the cubital fossa. The biceps tendon will stand out under your thumb.

- Instruct the patient to relax his or her arm. Maintain thumb over the patient's biceps tendon in the cubital fossa.

- Tap your thumbnail with the narrow end of a reflex mallet (see Figure 5.30). In a normal response, the patient's biceps should jerk slightly—a movement that you can both feel and see. The patient's fingers also will extend.

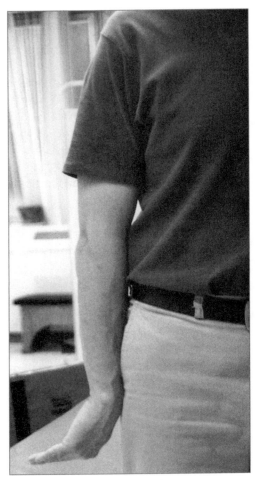

Figure 8.1 Waiter's Tip Position

☞ If the biceps tendon reflex is absent, upper brachial plexus damage may be suspected.

Brachioradialis tendon reflex screening:

- Place the patient's arm in 90° of elbow flexion across your forearm.

- Place hand under the medial aspect of the patient's elbow.

- Palpate the tendon of the brachioradialis at the distal end of the radius.

- Tap the brachioradialis tendon with the flat edge of a reflex mallet. In a normal response, the tap should elicit a small radial jerk (see Figure 5.31).

☞ If the brachioradialis and biceps reflexes are absent, upper brachial plexus damage can be suspected.

2. *Lower brachial plexus injury—Klumpke syndrome, clawhand deformity.* This disorder arises from C8 and T1 peripheral nerve roots, which innervate the muscles responsible for wrist and finger flexion and finger abduction and adduction, and occurs as a result of damage to the C8 and T1 roots. A forceful upward pull of the arm at birth may cause Klumpke syndrome. The presence of a cervical rib or a space-occupying lesion that compresses the lower part of the brachial plexus also can cause Klumpke syndrome. Lower plexus injuries can cause paralysis of all the intrinsic hand muscles and weakness of the wrist and finger flexors.

A clawhand deformity (see Figure 8.2) commonly is seen in lower plexus injuries. The digits are hyperextended at the metacarpalphalangeal (MCP) joints and flexed at the interphalangeal and distal interphalangeal (DIP) joints. The fifth digit remains abducted. Atrophy of the intrinsic muscles commonly cause the palmar surface to appear cupped or scooped away. Sensation is lost in the region of the ulnar side of the arm, forearm, and hand.

Lower plexus syndromes often are accompanied by sympathetic nervous system disturbances. Trophic changes in the affected arm may occur, including edema, dry skin, loss of hair, and hard thickened nails in the regions of denervation.

Figure 8.2 Clawhand Deformity

3. *Long thoracic nerve damage—winging of the scapula.* This disorder arises from C5, C6, and C7 peripheral nerve roots, which innervate muscles responsible for protraction and upward rotation of the scapula, and occurs from damage to the long thoracic nerve. The long thoracic nerve is prone to stretch or traction injuries (e.g., lifting heavy objects). Prolonged compression of the long thoracic nerve from lying on the lateral aspect of the trunk also can lead to damage of the nerve. Because the long thoracic nerve is trapped against the lower vertebrae, trauma to the base of the neck can cause damage to the nerve.

The long thoracic nerve innervates the serratus anterior muscle. When the nerve is damaged, winging of the scapula commonly occurs (see Figure 8.3). Instability of the scapula with medial rotation of the lower scapular region also can occur.

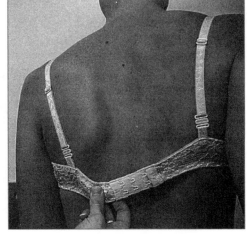

Figure 8.3 Winging of the Scapula

Wall push-up screen:

■ Instruct the patient to perform a wall push-up (causing protraction of the scapulae).

☞ If winging of the scapula is observed, long thoracic nerve damage may be indicated. Slight winging of the scapula at rest also may indicate damage to the long thoracic nerve.

4. *Median nerve damage—carpal tunnel syndrome, ape hand, benediction sign.* This disorder arises from C6, C7, C8, and T1 peripheral nerve roots, which innervate the muscles responsible for thumb opposition and abduction, forearm pronation, and wrist and finger flexion. The median nerve can become entrapped at several locations as it travels down the arm. Very often, entrapment of the median nerve is accompanied by entrapment of the ulnar nerve. Humeral fractures can disrupt median nerve signals at the elbow. Because the median nerve runs superficially at the wrist, it is vulnerable to damage resulting from wrist lacerations. Carpal bone injuries often result in medial nerve damage because the nerve travels directly over the carpal bones.

Carpal tunnel syndrome is the most common median nerve entrapment syndrome. The median nerve is compressed as it travels through the tunnel formed by the two rows of carpal bones and the flexor retinaculum (the fibrous ligamentous band that forms the carpal tunnel). In women, carpal tunnel syndrome may result from the hormonal changes occurring during pregnancy. Overuse syndromes and rheumatoid arthritis also may cause compression of the median nerve at the carpal tunnel.

Pain and paresthesias are present in the digits supplied by the median nerve. Such sensory disorders commonly become exacerbated at night. In severe cases, complete sensory loss may occur. Motor disorders include loss of thumb opposition, loss of the ability to make a fist, and atrophy of the thenar eminence.

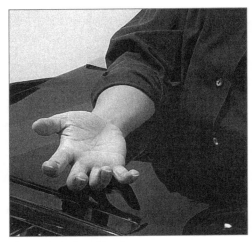

Figure 8.4 Ape Hand

Ape (simian) hand occurs as a result of denervation and its resultant atrophy of muscles of the thenar eminence. Thumb opposition is lost, and the hand flattens as a result of atrophy and paralysis (see Figure 8.4). *Benediction sign* occurs as a result of paralysis to the flexors of the third and fourth digits. When the patient is asked to make a fist, the third and fourth digits do not fully flex; instead, these fingers remain in a position similar to the benediction sign made by clergy (see Figure 8.5).

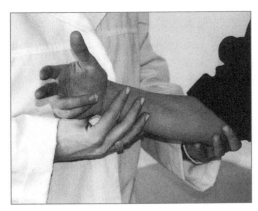

Figure 8.5 Benediction Sign

Tinel's sign for carpal tunnel syndrome:

- ■ Taps the patient's wrist over the carpal tunnel.

☞ If pain is felt in the dermatomal distribution of the median nerve, carpal tunnel syndrome can be suspected.

Phalen's test for carpal tunnel syndrome:

- ■ Maximally flex the patient's wrist and hold it in a flexed position for approximately 1 minute.

☞ If pain is felt in the distribution of the median nerve, carpal tunnel syndrome can be suspected.

5. *Ulnar nerve damage—cubital tunnel syndrome, Guyon's canal syndrome, bicycle rider's syndrome, clawhand deformity.* This disorder arises from C8 and T1 peripheral nerve roots, which innervate the muscles responsible for ulnar deviation of the wrist and fingers, thumb opposition and adduction, finger abduction and adduction, finger flexion at DIP joint (2nd–5th digits), and fifth digit opposition and abduction. The ulnar nerve can become entrapped and damaged at several locations as it travels down the arm. Very often, entrapment of the ulnar nerve is accompanied by entrapment of the median nerve. For example, inappropriate use of crutches can cause ulnar nerve compression when patients lean on crutches. The ulnar nerve in the axillary region can become compressed.

Cubital tunnel syndrome is the most common entrapment syndrome at the elbow. Elbow deformities, repetitive motion at the elbow, inflammatory conditions, and trauma to the elbow all may cause cubital tunnel entrapment.

The ulnar nerve can be entrapped as it travels from the forearm into the hand through *Guyon's canal* (the ulnar tunnel). The floor of the tunnel is formed by the pisiform and hamate bones; the upper portion of the tunnel is formed by the pisohamate ligament. Overuse of or damage to the hypothenar eminence of the hand often results in compression of Guyon's canal.

Bicycle rider's syndrome is compression of the ulnar nerve in the hypothenar eminence caused by prolonged bicycle riding. Similarly, inappropriately pressing the hypothenar eminence on a crutch can result in ulnar nerve compression.

Motor deficits at the wrist can present as radial deviation of the wrist. This occurs as a result of denervation of the flexor carpi ulnaris muscle. Pain is a common feature of ulnar nerve entrapment syndromes; however, sensory deficits also can include hyperesthesia; paresthesia; and, in severe cases, complete sensory loss.

Clawhand deformity occurs as a result of ulnar nerve damage. The first phalanx is hyperextended, and the distal two phalanges are flexed; the fifth digit is abducted; there is an inability to voluntarily abduct or adduct the fingers; weakened thumb opposition and total loss of fifth finger opposition occur (see Figure 8.2); and atrophy of the intrinsic muscles of the fingers and flattening of the hypothenar eminence occur.

Froment's sign:

- Instruct the patient to grasp a piece of paper between the thumb and the index finger.

Because the adductor pollicis muscle is denervated, the patient is unable to adduct the thumb and instead flexes it. Flexion of the thumb becomes more pronounced when the therapist pulls the paper away.

6. *Radial nerve damage—Saturday night palsy, tennis elbow, wrist drop.* This disorder arises from C5, C6, C7, C8, and T1 peripheral nerve roots, which innervate the muscles responsible for radial deviation of the wrist and fingers; extension of the shoulder, elbow, wrist, and fingers; elbow flexion; forearm supination and pronation; and thumb abduction at the carpal metacarpal (CMC) joint.

The radial nerve is particularly vulnerable to compression/entrapment as a result of its location in the brachial plexus and its proximity to the humerus. For example, inappropriate use of crutches can place pressure on the nerve at the spiral groove, causing radial nerve damage. The radial nerve is also vulnerable to trauma because it runs superficially during part of its course and lies against the rigid spiral groove of the humerus. Shoulder dislocation, humeral fractures, and radial neck fractures can all damage the radial nerve. Because of the nerve's superficial location at part of its course, the radial nerve often is damaged as a result of gunshot and stab wounds.

The radial nerve is commonly damaged as a result of pressure placed on the nerve at the spiral groove of the humerus. *Saturday night palsy* results from compression of the radial nerve at the spiral groove. When an individual who is inebriated falls asleep with an arm against a rigid surface (e.g., a park bench or sidewalk) and all muscles innervated by the radial nerve become denervated (except the triceps muscle), Saturday night palsy occurs.

Compression of a branch of the radial nerve at the lateral epicondyle of the humerus commonly results in *tennis elbow*—a severe form of pain experienced on the lateral side of the elbow. Overuse syndromes (e.g., repetitive extension of the elbow during tennis) can result in radial nerve damage.

Most lesions to the radial nerve affect all muscles innervated by the radial nerve except the triceps. Such lesions usually produce an inability to extend the MCP joints, the wrist, and the thumb; inability to supinate (unless the biceps muscle is used); weakness in thumb abduction (although thumb opposition remains intact); and paralysis of the brachioradialis muscle, resulting in weak elbow flexors. Pain and paresthesias are present in the distribution of the radial nerve. In severe cases, complete sensory loss can occur.

Wrist drop is a common sign of radial nerve damage due to denervation of the wrist extensors (see Figure 8.6). Impaired gripping also is commonly observed. Patients have difficulty gripping objects as a result of an inability to extend their wrists.

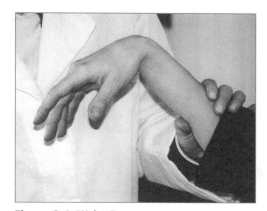

Palm-to-palm screen:

- Place the patient's hands palm-to-palm (in a prayer position).

- Instruct the patient to separate his or her hands.

Figure 8.6 Wrist Drop

☞ Wrist drop will occur in the affected side if the radial nerve has been damaged.

Grip screen:

- Present the patient with a tennis ball (or a ball the size of a tennis ball) and instruct the patient to grasp the ball with his or her affected side.

☞ If the patient has difficulty extending his or her wrists and fingers to grasp the ball, radial nerve damage can be suspected.

COMMON PERIPHERAL NERVE SYNDROMES OF THE LOWER EXTREMITY

Lumbar plexus disorders. The lumbar plexus arises from L1, L2, and L3, with a contribution from L4. True lesions of the lumbar plexus are rare because of the plexus's location deep within the abdomen. Damage to the plexus often is accompanied by fatal injuries.

Fractures and space-occupying lesions may occasionally damage the plexus.

1. *Femoral nerve damage—knee buckling, loss of patellar reflex.* This disorder arises from L2, L3, and L4 peripheral nerve roots, which innervate the muscles responsible for hip flexion and knee extension. The femoral nerve is the most commonly damaged peripheral nerve of the lumbar plexus; it is vulnerable to compression as it travels through the pelvis. Tumors of the vertebrae and fractures of the pelvis and upper femur can damage the femoral nerve. Femoral neuropathies also may result from vascular compromise secondary to diabetes.

If damage to the femoral nerve causes the iliacus muscle to be denervated, weakness in hip flexion will result. The nerve to the quadriceps muscle is the most commonly injured branch of the femoral nerve. Loss of innervation of the quadriceps causes difficulty walking; the patient's knee frequently buckles. Walking stairs becomes particularly difficult as a result of a denervated quadriceps muscle. Lesions to the sensory portions of the femoral nerve result in diminished or lost sensation of the anterior thigh and the medial leg and foot. Often, the patient reports radiating pain in the medial aspect of the knee; the pain radiates distally to the medial aspect of the involved foot.

Patellar tendon reflex screen:

- Ask the patient to sit with knees dangling (or with one leg crossed over the other).

- Palpate the soft tissue depression on either side of the tendon in order to locate it accurately and attempt to elicit the reflex by tapping the tendon with the flat edge of a reflex mallet at the level of the knee joint. A normal reflex is displayed by knee extension.

☞ If the reflex is absent, femoral nerve damage is indicated.

Squat-to-stand screen:

- Instruct the patient to stand upright from a squatting position.

- Note whether the patient is able to stand straight, with knees in full extension, or whether the patient uses one leg more than the other. The arc of motion from flexion to extension should be smooth.

Occasionally, the patient may be able to extend the knee smoothly only until the last 10°, finishing the motion with great effort. This faltering in the last 10° of extension is called *extension lag*; it occurs because the last 10° to 15° of knee extension requires more muscle power than the rest.

Walking stairs screen:
The patient experiences great difficulty walking up and down stairs as a result of denervation of the quadriceps muscle.

Sacral plexus disorders. The sacral plexus is formed by the lumbosacral trunk, which arises from the lumbar plexus (L4) and the ventral roots of L5, S1, S2, and S3. Contributions also may come from S4. Injuries to the sacral plexus are rare; however, damage to the roots of the lumbar region occur frequently as a result of disk disease or disorder. Fractures and space-occupying lesions may sometimes cause damage to the sacral plexus.

1. *Sciatic nerve damage—sciatica, foot drop (with Achilles tendon reflex absent).* This disorder arises from L4, L5, S1, S2, and S3 peripheral nerve roots, which innervate the muscles responsible for hip extension and adduction; knee flexion; plantar and dorsiflexion; foot inversion and eversion; and toe extension, flexion, abduction, and adduction. Lesions to the sciatic nerve are always accompanied by loss of function of the common peroneal nerve, the tibial nerve, or both. Sciatic nerve damage can result from a variety of causes:

- Intramuscular injections improperly administered in the buttock region

- Fractures of the pelvis and femur

- Tumors, stab and gunshot wounds, and prolonged squatting

- Piriformis syndrome, which occurs when the sciatic nerve is compressed between the piriformis and the obturator internus muscles as the nerve exits the pelvis in the greater sciatic foramen. Referred to as *sciatica*.

Loss of the sciatic nerve results in loss of voluntary knee flexion secondary to denervation of the hamstring muscles. Diminished innervation of all the muscles of the leg and foot results in a steppage gait (the compensation for a lack of dorsiflexion by using excessive hip flexion) and an inability to run. Foot drop (see Figure 8.7) is observable.

Sensory deficits include loss of sensation on the lateral side of the leg and foot. Pain and/or diminished sensation is experienced on the posterior aspect of the leg and the plantar surface of the foot (sciatica). Both the Achilles and plantar reflexes are lost, there is an inability to stand on the toes or heels, and proprioception is lost in the foot and toes.

Foot drop results due to denervation of the dorsiflexors.

Achilles tendon reflex screen:

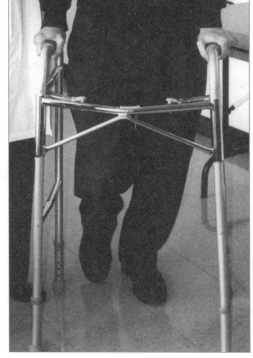

Figure 8.7 Footdrop

- Ask the patient to sit with legs dangling. If the patient is lying down supine, flex one leg at both the hip and knee; rotate the hip internally so that the leg rests across the opposite shin (see Figure 5.34).

- Place thumb and fingers into the soft tissue depressions on either side of the tendon to locate the patient's Achilles tendon accurately.

- Strike the tendon with the flat end of a reflex mallet to induce a sudden, involuntary plantar flexion of the foot.

☞ The absence of the Achilles tendon reflex—along with foot drop—often indicates sciatic nerve damage.

Plantar reflex screen:

- Place the patient in supine.

- Use a firm object (e.g., a key or a wooden tongue depressor) to stroke the patient's foot on the plantar surface. Stroke the foot on the lateral aspect, starting from the heel and moving upward to the ball of the foot. At the ball of the foot begin curving across the foot to the medial aspect (see Figure 5.35). A normal response is indicated by toe flexion.

☞ The absence of both the Achilles tendon and the plantar reflex indicates sciatic nerve damage.

Toe and heel stand screen:

- Instruct the patient to stand on his or her toes and heels.

☞ Inability to perform these movements indicates sciatic nerve damage.

Lasegue's sign:

- Place the patient's sciatic nerve on stretch by actively straight raising the patient's leg (while supine).

☞ Pain along the sciatic nerve distribution indicates sciatic nerve damage.

Proprioception screening:

- With the patient's eyes closed, actively position the involved foot (e.g., plantar flexion and eversion).

- Ask the patient to mirror the position with his or her uninvolved foot.

☞ If the patient is unable to detect the position of the involved foot, proprioception is impaired. Sciatic nerve damage can be suspected.

2. *Tibial nerve damage—tarsal tunnel syndrome.* This disorder arises from L4, L5, S1, S2, and S3 peripheral nerve roots, which innervate the muscles responsible for knee flexion; plantar flexion; foot inversion; and toe flexion, abduction, and adduction. The tibial nerve arises from the sciatic nerve above the level of the popliteal fossa. Damage to the sciatic nerve often involves nerve portions that form the tibial nerve. Isolated damage to the tibial nerve commonly is due to an injury in or below the popliteal space—a region that is highly vulnerable to trauma. *Tarsal tunnel syndrome* occurs when the tibial nerve is compressed by the flexor retinaculum (the tarsal tunnel) as the nerve passes behind and beneath the medial malleolus.

Damage to the tibial nerve results in a loss of plantar flexion; foot adduction and inversion; and toe flexion, abduction, and adduction. Patients are unable to stand on their toes and report that walking is both fatiguing and painful. The Achilles tendon reflex also will be absent. Tarsal tunnel syndrome can result in pain and sensory loss on the plantar surface of the foot.

Achilles tendon reflex screen:

- Ask the patient to sit with legs dangling. If the patient is lying supine, flex one leg at both the hip and knee; rotate the hip internally so that the leg rests across the opposite shin (see Figure 5.34).

- Place thumb and fingers into the soft tissue depressions on either side of the tendon to locate the patient's Achilles tendon accurately.

- Strike the tendon with the flat end of a reflex mallet to induce a sudden, involuntary plantar flexion of the foot.

☞ The absence of the Achilles tendon reflex may indicate tibial nerve damage.

Tarsal tunnel syndrome screen:

- ■ Tap the medial malleolus just above the margin of the flexor retinaculum.

☞ Tarsal tunnel syndrome is indicated by paresthesias felt in the foot.

3. *Common peroneal nerve—foot drop (with Achilles tendon reflex preserved).* This disorder arises from L4, L5, S1, and S2 peripheral nerve roots, which innervate the muscles responsible for dorsiflexion, foot eversion, and toe extension. The common peroneal nerve arises from the sciatic nerve above the level of the popliteal fossa. Damage to the sciatic nerve usually involves nerve portions that form the common peroneal nerve. The common peroneal nerve is vulnerable to damage because of its superficial course as it travels by the head of the fibula. Because the nerve is firmly attached to the fibular head, it cannot move when compressed and, thus, is easily damaged. Prolonged sitting in a cross-legged position or squatting can compress the common peroneal nerve at the fibula neck. The nerve also may be injured at this site during sleep. Patients who are bedridden and allowed to maintain their legs in excessive internal rotation are prone to compression injuries of the common peroneal nerve.

Lesions to the common peroneal nerve can result in an inability to dorsiflex or evert the foot. To compensate for a lack of dorsiflexion, patients use excessive hip flexion, resulting in a steppage gait. Foot drop commonly is observed (see Figure 8.7). Sensation often is lost or diminished on the lateral portion of the leg and on the dorsum of the foot. Pain rarely is felt as a result of common peroneal nerve damage.

Lesions to the common peroneal nerve can be differentiated from sciatic nerve lesions. In common peroneal nerve damage, the Achilles tendon reflex is preserved. In sciatic nerve lesions, the Achilles tendon reflex is absent.

- ■ Ask the patient to dorsiflex and evert the involved foot.

☞ If the patient is unable to perform these movements, common peroneal nerve damage may be suspected.

Achilles tendon reflex screen:

- ■ Ask the patient to sit with legs dangling. If the patient is lying supine, flex one leg at both the hip and knee; rotate the hip internally so that the leg rests across the opposite shin (see Figure 5.34).

- ■ Place his or her thumb and fingers into the soft tissue depressions on either side of the tendon to locate the patient's Achilles tendon accurately.

- ■ Strike the tendon with the flat end of a reflex mallet to induce a sudden, involuntary plantar flexion of the foot.

☞ If the reflex is intact but foot drop is observable, common peroneal nerve damage is indicated. (If foot drop is observable but the Achilles tendon reflex is absent, sciatic nerve involvement is indicated.)

RED FLAGS
SIGNS AND SYMPTOMS OF NEUROPATHY
OR PERIPHERAL NERVE DAMAGE

A red flag is an indicator or symptom of dysfunction. If a patient is observed displaying any of the following red flags, it is likely that the patient possesses neuropathy or peripheral nerve damage.

- **STOCKING-AND-GLOVE NEUROPATHY.** The *distal segments* of an extremity are involved before the *proximal segments*. The sensation of the feet and knees are affected first, followed by the hands. Stocking-and-glove neuropathy is representative of nerve damage that follows a *peripheral distribution*. (If muscular dysfunction is indicated rather than nerve damage, the proximal segments of an extremity are affected first, followed by the distal extremities.)

- **PARESTHESIA.** The sensation of pins and needles.

- **CAUSALGIA.** Burning sensation.

- **DYSESTHESIA.** Pain.

- **AUTONOMIC DYSFUNCTION.** The skin over the denervated area appears smooth and glossy; the nails become thick; hair thins or falls out; alterations in sweating occur (also referred to as *trophic changes*).

- **HYPERESTHESIA.** Heightened sensation over a denervated area.

- **FLACCIDITY,** resulting from complete denervation. Muscular weakness occurs as a result of partial denervation, causing the patient to fatigue quickly.

- **HYPOREFLEXIVE DEEP TENDON REFLEXES.** Respond sluggishly when elicited on examination.

- **REFLEX SYMPATHETHIC DYSTROPHY (RSD),** or frozen shoulder. RSD is characterized by immobility and pain in an upper extremity having nerve damage.

- **UPPER BRACHIAL PLEXUS DAMAGE.** Waiter's tip position, absent biceps and brachioradialis reflexes, diminished or lost sensation on the deltoid and radial surfaces of the forearm and hand.

- **LOWER BRACHIAL PLEXUS DAMAGE.** Clawhand deformity; diminished or lost sensation on the ulnar side of the arm, forearm, and hand.

- **LONG THORACIC NERVE DAMAGE.** Winging of the scapula.

- **MEDIAN NERVE DAMAGE.** Carpal tunnel syndrome, ape hand, benediction sign, lost thumb opposition, lost ability to make a fist, atrophy of thenar eminence, pain and paresthesias present in the third and fourth digits.

- **ULNAR NERVE DAMAGE.** Cubital tunnel syndrome, Guyon's canal syndrome, bicycle rider's syndrome, clawhand deformity, atrophy of intrinsic muscles of the fingers, flattening of the hypothenar eminence, pain and paresthesias along the ulnar nerve distribution.

- **RADIAL NERVE DAMAGE.** Saturday night palsy, tennis elbow, wrist drop, impaired gripping, pain and paresthesias in the radial distribution.

- **FEMORAL NERVE DAMAGE.** Weakness in hip flexion, knee buckling, difficulty walking steps, lost patellar reflex, diminished/lost sensation to the anterior thigh, and medial leg and foot.

- **SCIATIC NERVE DAMAGE.** Sciatica, foot drop, absent Achilles tendon reflex and patellar reflex, lost voluntary knee flexion, steppage gait (use of excessive hip flexion), inability to stand on toes or heels, diminished or lost sensation on the lateral side of the leg and foot, pain or diminished sensation on the posterior aspect of the leg and plantar surface of the foot.

- **TIBIAL NERVE DAMAGE.** Tarsal tunnel syndrome; lost plantar flexion, foot adduction and inversion, and toe flexion, abduction, and adduction; inability to stand on toes or heels; absent Achilles tendon reflex; pain or sensory loss on the plantar surface.

- **COMMON PERONEAL NERVE DAMAGE.** Foot drop (with Achilles tendon reflex preserved), inability to dorsiflex or evert the foot, steppage gait (use of excessive hip flexion), diminished or lost sensation on the lateral portion of the leg and on the dorsum of the foot (pain rarely is felt).

AVAILABLE IN-DEPTH ASSESSMENTS

An in-depth assessment for neuropathy and peripheral nerve damage can be administered and interpreted only by a physician (i.e., a neurologist). The most common in-depth assessment is EMG. EMG can demonstrate the presence or absence of normal innervation to a muscle by evaluating the muscle's nerve conduction and electrical activity.

Therapists who wish to use an upper-extremity sensorimotor assessment to further determine how an upper-extremity peripheral neuropathy or injury has affected the patient's function can use the following in-depth assessments:

Box and Block Test of Manual Dexterity
V. Mathiowetz, G. Volland, N. Kashman, K. Weber
Adults with fine motor deficits secondary to neurologic and/or musculoskeletal impairment
Mathiowetz, V., Volland, G., Kashman, N., & Weber, K. (1985). Adult norms for the Box and Block Test of Manual Dexterity. *American Journal of Occupational Therapy, 39,* 386–391.

Jebsen Hand Function Test
R. H. Jebsen, N. Taylor, R. B. Trieschmann, M. J. Trotter, L. A. Howard

Children and adults with neurologic and/or musculoskeletal impairment
Sammons Preston Inc.
PO Box 50710
Bolingbrook, IL 60440-5071
(800) 323-5547

Minnesota Rate of Manipulation Tests (MRMT) and Minnesota Manual Dexterity Test (MMDT)
University of Minnesota Employment Stabilization Research Institute
Adults with neurologic and/or musculoskeletal impairment
American Guidance Services, Inc.
Publishers' Building
Circle Pines, MN 55014

Nine-Hole Peg Test
V. Mathiowetz, K. Weber, N. Kashman, G. Volland
Adults with fine motor coordination deficits secondary to neurologic and/or musculoskeletal impairment
Mathiowetz, V., Weber, K., Kashman, N., & Volland, G. (1985). Adult norms for the Nine-Hole Peg Test of finger dexterity. *Occupational Therapy Journal of Research, 5,* 24–38.

O'Connor Finger and Tweezer Dexterity Tests

J. O'Connor

Adults with fine motor coordination deficits secondary to neurologic and/or musculoskeletal impairment

Sammons Preston Inc.
PO Box 5071
Bolingbrook, IL 60440-5071
(800) 323-5547

Purdue Pegboard

J. Tiffin

Adults with fine motor and eye–hand coordination deficits secondary to neurologic and/or musculoskeletal impairment

Lafayette Instrument Company
3700 Sagamore Pkwy. North
PO Box 5729
Lafayette, IN 47903

Semmes–Weinstein Monofilaments

J. Semmes, S. Weinstein

Adults with sensory problems

Sammons Preston Inc.
PO Box 5071
Bolingbrook, IL 60440-5071
(800) 323-5547

Therapists who wish to use an in-depth balance assessment to further determine how a lower-extremity peripheral neuropathy or injury has affected the patient's function and risk for falling can use the following in-depth assessments:

Activities-Specific Balance Confidence Scale (ABC)

L. E. Powell, A. M. Myers

Adults with balance deficits

Powell, L. E., & Myers, A. M. (1995). The Activities-Specific Balance Confidence (ABC) Scale. *Journal of Gerontology, 50A*(1), M28–M34.

Berg Balance Scale (Berg)

K. O. Berg, S. L. Wood-Dauphinee, J. I. Williams, B. Maki

Adults with balance deficits

Berg, K. O., Wood-Dauphinee, S. L., Williams, J. I., & Maki, B. (1992). Measuring balance in the elderly: Validation of an instrument. *Canadian Journal of Public Health, 83*, S7–S11.

Clinical Test of Sensory Interaction in Balance (CTSIB)

M. Watson

Adults with balance deficits

Watson, M. (1992). Clinical Test of Sensory Interaction in Balance. *Physiotherapy Theory and Practice, 8*(4), 176–178.

Falls Efficacy Scale (FES)

M. E. Tinnetti, D. Richman, L. Powell

Elderly persons with balance deficits

Tinnetti, M. E., Richman, D., & Powell, L. (1990). Falls efficacy as a measure of fear of falling. *Journal of Gerontology, 45*, 239–243.

Functional Reach Test (FRT)

P. W. Duncan, D. K. Weiner, J. Chandler, S. A. Studenski

Adults with balance deficits secondary to neurologic damage

Duncan, P. W., Weiner, D. K., Chandler, J., & Studenski, S. A. (1990). Functional reach: A new clinical measure of balance. *Journal of Gerontology, 45*, 192–197.

Morse Fall Scale (MFS)

J. M. Morse

Adults with balance deficits

Janice M. Morse
Pennsylvania State University
School of Nursing
201 Health and Human Development East
University Park, PA 16802-6508

Physical Performance Test (PPT)

D. B. Reuben, A. L. Siu

Elderly persons with balance deficits

Reuben, D. B., & Siu, A. L. (1990). An objective measure of physical function of elderly outpatients: The Physical Performance Test. *Journal of the American Geriatric Society, 38*, 1105–1112.

Timed Get Up and Go (TGUG)
D. Podsiadlo, S. Richardson
Adults with balance deficits
Podsiadlo, D., & Richardson, S. (1991). The Timed Get Up and Go: A test of basic functional mobility for frail elderly persons. *Journal of the American Geriatric Society, 39,* 142–148.

Tinnetti Balance Test of the Performance-Oriented Assessment of Mobility Problems (Tinnetti)
M. E. Tinnetti
Adults with balance deficits
Tinnetti, M. E. (1986). Performance-oriented assessment of mobility in elderly patients. *Journal of the American Geriatric Society, 34,* 119–126.

For further information about the these assessments, refer to the following resources:

Asher, I. E. (1996). *Occupational therapy assessment tools: An annotated index* (2nd ed.). Bethesda, MD: American Occupational Therapy Association.

Impara, J. C., Plake, B. S., & Murphy L. L. (Eds.). (1998). *The thirteenth mental measurements yearbook.* Lincoln: University of Nebraska Press.

Murphy, L. L., Impara, J. C., & Plake, B. S. (Eds.). (1999). *Tests in print: An index to tests, test reviews, and the literature on specific tests* (Vols. 1 & 2). Lincoln: University of Nebraska Press.

Plake, B. S., Impara, J. C., & Murphy, L. L. (Eds.). (1999). *The supplement to* The Thirteenth Mental Measurements Yearbook. Lincoln: University of Nebraska Press.

SCREENING FORM FOR NEUROPATHY AND PERIPHERAL NERVE DAMAGE

Patient Name _____ Date _____

SENSATION

Dermatomal area of impairment:

Presence of sensory phenomenon:

Paresthesia (pins and needles)	☐ Yes	☐ No
Dysesthesia (pain)	☐ Yes	☐ No
Causalgia (burning sensation)	☐ Yes	☐ No
Hyperesthesia (heightened sensation)	☐ Yes	☐ No

Presence of autonomic dysfunction and trophic skin changes on the denervated region(s):

Glossy, smooth skin	☐ Yes	☐ No
Thickened finger/toe nails	☐ Yes	☐ No
Thinning and/or loss of hair	☐ Yes	☐ No

MOTOR

Presence of impairment found on manual muscle testing:

Involved body part and action: Muscle grade:

Muscular weakness	☐ Yes	☐ No
Flaccidity	☐ Yes	☐ No
Muscular atrophy	☐ Yes	☐ No
Rapidly fatiguing muscle	☐ Yes	☐ No
Muscular cramping	☐ Yes	☐ No

Presence of hyporeflexive/areflexive deep tendon reflexes:

Biceps tendon	☐ Absent	☐ Hyporeflexive	☐ Normal
Brachioradialis tendon	☐ Absent	☐ Hyporeflexive	☐ Normal
Patellar tendon	☐ Absent	☐ Hyporeflexive	☐ Normal
Achilles tendon	☐ Absent	☐ Hyporeflexive	☐ Normal
Plantar tendon	☐ Absent	☐ Hyporeflexive	☐ Normal

Presence of impairment found on ROM test:

ROM limitations:

Involved joint: ROM:

SOFT TISSUE INTEGRITY

Presence of impairment found on soft tissue examination:

Skin breakdown	☐ Yes	☐ No
Edema	☐ Yes	☐ No
Adhesions and/or fascia thickening	☐ Yes	☐ No
Joint instability/weakness/hypermobility	☐ Yes	☐ No
Subluxation	☐ Yes	☐ No
Damaged articular surfaces 2° to joint capsule deterioration (bony end-feels)	☐ Yes	☐ No

CLINICAL TESTS

Brachial plexus damage:

Deep tendon reflexes:			
Biceps tendon reflex	☐ Absent	☐ Hyporeflexive	☐ Normal
Brachioradialis tendon reflex	☐ Absent	☐ Hyporeflexive	☐ Normal

Long thoracic nerve damage:		
Wall push-up test: Winging of the scapula	☐ Yes	☐ No

Median nerve damage:		
Tinel's sign for carpal tunnel syndrome (Is pain elicited by tapping the wrist over the carpal tunnel?)	☐ Yes	☐ No
Phalen's test for carpal tunnel syndrome (Is pain elicited by maximally flexing the wrist for 1 minute?)	☐ Yes	☐ No
Is ape (simian) hand observable?	☐ Yes	☐ No
Is atrophy of the thenar eminence observable?	☐ Yes	☐ No
Is the benediction sign observable?	☐ Yes	☐ No

Ulnar nerve damage:		
Is Froment's sign (thumb flexion, rather than adduction, becomes pronounced when the patient attempts to grasp a piece of paper between the thumb and index finger) visible?	☐ Yes	☐ No
Is clawhand deformity observable?	☐ Yes	☐ No
Is atrophy of the hypothenar eminence observable?	☐ Yes	☐ No

Radial nerve damage:		
Palm-to-palm test (Does wrist drop occur on the affected side when the patient separates both hands from a prayer position?)	☐ Yes	☐ No
Grip test (Does the patient have difficulty extending the affected wrist and fingers when asked to grasp a tennis ball?)	☐ Yes	☐ No
Does the patient experience severe pain on the lateral side of the affected elbow (tennis elbow)?	☐ Yes	☐ No

Lumbar plexus damage:

Femoral nerve damage:			
Patellar tendon reflex	☐ Absent	☐ Hyporeflexive	☐ Normal
Squat-to-stand test (Is the patient able to stand upright from a squatting position with knees in full extension?)	☐ Yes	☐ No	
Walking stairs (Is the patient able to walk up and down stairs?)	☐ Yes	☐ No	
Is knee buckling observable?	☐ Yes	☐ No	

Sacral plexus damage:

Sciatic nerve damage:			
Achilles tendon reflex	☐ Absent	☐ Hyporeflexive	☐ Normal
Plantar tendon reflex	☐ Absent	☐ Hyporeflexive	☐ Normal
Toe and heel stand (Is the patient able to stand on his toes and heels?)	☐ Yes	☐ No	
Lasegue's sign (Does the patient experience pain along the sciatic distribution when the affected leg is passively raised, with patient in supine?)	☐ Yes	☐ No	

Proprioception testing of the involved foot and toes	☐ Intact	☐ Impaired	
Is a steppage gait observable?	☐ Yes	☐ No	
Is foot drop observable?	☐ Yes	☐ No	
Tibial nerve damage:			
Achilles tendon reflex	☐ Absent	☐ Hyporeflexive	☐ Normal
Tarsal tunnel syndrome (Are paresthesias felt in the foot when the patient's medial malleolus is tapped just above the margin of the flexor retinaculum?)	☐ Yes	☐ No	
Common peroneal nerve damage:			
Achilles tendon reflex	☐ Absent	☐ Hyporeflexive	☐ Normal
Is foot drop observable?	☐ Yes	☐ No	
Does the patient exhibit a steppage gait?	☐ Yes	☐ No	

Dysphagia
SCREENING

FUNCTIONAL IMPLICATIONS OF SWALLOWING IMPAIRMENT

Swallowing is divided into four stages: (a) the oral preparatory stage, (b) the oral stage, (c) the pharyngeal stage, and (d) the esophageal stage. The *oral preparatory stage* involves preparing the food for transport through the oral cavity, including bolus manipulation, mastication, and formation. This stage requires coordinated movements of the lips, jaw, cheeks, and soft palate. Factors such as bolus volume, viscosity, and temperature are perceived and account for muscular adjustments required for appropriate preparation. The oral preparatory stage of the swallow requires adequate buccal tone, labial tone, and mandibular and lingual range of motion (ROM).

The *oral stage* of the swallow consists of lingual propulsion of the food through the mouth. This stage ends when the pharyngeal swallow is triggered. Adequate lingual ROM and coordination are necessary for movement of food through the oral cavity.

The *pharyngeal stage* begins when the swallow is triggered and continues until the bolus passes through the upper esophageal sphincter. During the pharyngeal stage of the swallow, certain physiological activities occur, including

- Velar elevation to prevent nasal regurgitation

- Laryngeal closure to prevent material from entering the airway

- Tongue base contact with the posterior pharyngeal wall

- Contraction of the pharyngeal constrictors to allow for movement of the food through the pharynx

- Relaxation of the cricopharyngeal sphincter to allow material to pass into the esophagus.

The larynx elevates superiorly and anteriorly in order to protect the airway. The sequence of events that allow for laryngeal closure include

- True vocal fold closure

- False vocal fold closure

- Anterior and medial movement of the aryepiglottic folds

- Epiglottic retroversion.

The *esophageal stage* occurs when peristalsis moves the bolus through the esophagus into the stomach.

The oral preparatory and oral stages of the swallow are regulated by cranial nerves 5, 7, and 12. Patients with *cranial nerve 5 involvement* (trigeminal nerve—motor) may experience decreased mastication/chewing abilities. Patients with *cranial nerve 7 involvement* (facial nerve—motor) may experience difficulty controlling food and liquid in the mouth, pocketing of food in the cheeks, and spillage of material from the oral cavity secondary to reduced lip closure. Patients with *cranial nerve 12 involvement* (hypoglossal nerve) may experience decreased bolus control because of reduced lingual movement and coordination.

The pharyngeal stage of the swallow can be negatively affected by damage to cranial nerves 9 and 10 (sensory and motor branches). Patients with *cranial nerve 9 involvement* (glossopharyngeal nerve—sensory) may experience a delayed triggering of the pharyngeal swallow and possibly subsequent spillage of material into the airway before the initiation of the swallow. Patients with *cranial nerve 10 sensory branch involvement* (superior laryngeal nerve) may display a loss of laryngeal closure and a cough reflex. Patients with *cranial nerve 10 motor branch involvement* (vagus) may display inadequate velopharyngeal closure, hypopharyngeal residue, and inadequate laryngeal closure.

DISORDERS AND SWALLOWING PATTERNS

In general, patients with certain disorders present with classic patterns of dysphagia. The most common disorders and their swallowing patterns are outlined here.

Cerebrovascular Accident (CVA)

CVA usually results in *unilateral weakness* of the tongue, lip, face, and laryngeal and pharyngeal areas, subsequently affecting coordination and movement. Swallowing deficits observed during a screening may include

- Difficulty stripping the food from the utensil

- Difficulty containing the food or liquid in the oral cavity

- Decreased mastication

- Reduced oral transport

- Pocketing on one side of the cheek.

A patient with a *right hemisphere CVA* usually displays a rapid rate of intake secondary to decreased cognitive functioning (i.e., decreased attention, poor judgment). A patient with a *left hemisphere CVA* may have an *oral apraxia,* which may make initiation of the oral stage of the swallow difficult.

Parkinson's Disease

Parkinson's disease usually results in limited range of oral movements, tremors, and rigidity. Swallowing impairments observed during a screening may include

- Anterior loss of food or liquid from the oral cavity

- Decreased mastication

- Poor oral transport secondary to irregular tongue movements

- Residue in the oral cavity after the swallow.

Traumatic Brain Injury (TBI)

TBI may result in oral motor weakness and oral apraxias. Cognitive impairments (i.e., decreased attention, poor judgment, decreased awareness) may affect swallowing function, especially the oral preparatory stage of the swallow. Swallowing impairments observed during a screening may include

- Rapid rate of intake

- Large amounts taken on the spoon; the patient may stuff food into his or her mouth

- Unawareness of residue or pocketing, thereby a failure to reswallow in order to clear oral residual

- Reduced oral transport, especially if tongue movements are limited.

Alzheimer's Disease

Dysphagia may be caused by neuromuscular weakness, may be secondary to a behavioral pattern, or both. A patient with Alzheimer's disease may have reduced lingual ROM and coordination. Swallowing impairments observed during a screening may include

- Rapid rate and large amounts of intake

- Forgetting how to eat or that one is eating

- Vocalization while eating

- Poor bolus manipulation and oral transport.

DYSPHAGIA SCREENING PROCEDURES

A dysphagia screening should be performed on all patients with CVA, TBI, progressive neurological conditions (i.e., Parkinson's disease, amyotrophic lateral sclerosis, multiple sclerosis, myasthenia gravis), cerebral palsy, Alzheimer's disease, and head and neck cancer. The major purposes of a dysphagia screening are to determine

- The anatomic and functional status of the oral structures

- Whether the patient is at risk for aspiration on his or her current diet

- Whether further evaluation is warranted

- Whether the patient is an appropriate candidate for an instrumental examination.

The therapist must understand that a dysphagia screening is not an in-depth evaluation. It cannot detect "silent" aspiration, which studies have shown is present in approximately half of patients with dysphagia. In addition, screening cannot tell the therapist why aspiration or penetration may be occurring.

The following information should be obtained through a thorough review of the medical chart:

- *Diagnosis.* The patient's medical diagnosis is important to ascertain because certain diseases are innately characterized by dysphagia. Knowing the diagnosis allows the therapist to deduce typical signs that may accompany that diagnosis. For example, a person with a left CVA usually has a right facial weakness, which may lead to pocketing of material in the right side of the cheek. A person with Parkinson's disease usually has uncoordinated lingual movements, which may result in decreased bolus manipulation and transfer.

- *Medical history.* A patient's medical history provides insight into the possibility that the presence of previous medical conditions may either exacerbate or contribute to the present medical condition underlying the dysphagia.

- *Medications.* Medications may induce a dysphagia secondary to the side effect of the drug. For example, certain psychotropic drugs may lead to xerostomia (dry mouth), which can impair the oral stage of the swallow.

- *Current feeding status.* A therapist must be aware of the patient's current feeding status in order to determine whether it is functional, appropriate, and practical to the patient's health, nutritional, and safety requirements.

A basic cognitive screening should be performed before the dysphagia screening and include examination of yes/no reliability. This examination provides information regarding whether the patient is a reliable informant and whether he or she is a candidate for further instrumental testing and therapy. Interview the patient and family to determine the duration of the problem, frequency of occurrence, and factors that exacerbate or relieve the problem. It is also important to find out what type of foods the patient was eating before admission. Ideally, the patient should be observed during mealtime in a naturalistic environment (i.e., during dinner or lunch) in order to get a true sense of mealtime habits and behaviors. Patients should wear dentures and glasses if they normally do so during a meal in order to observe their usual state of eating.

Note: A dysphagia screening should include an oral peripheral examination, laryngeal examination, and observation with food trials.

Oral Peripheral Examination

The purpose of an oral peripheral examination is to assess oral structures and their functional movements. An oral peripheral exam allows the therapist to make a decision, based on the integrity and function of these structures, regarding the safety of subsequent feeding trials. The results of the oral–motor examination aid the clinician in making an informed decision about the safety of food administration during a swallowing screening.

1. *Labial ROM.*

- Ask the patient to pucker his or her lips and then retract them into a smile with lips closed.

- Note whether one side of the lips has greater ROM than the opposite side.

- Movements should be coordinated and symmetrical. The lips should appear closed and symmetrical at rest.

☛ Movements that are uncoordinated and asymmetrical may indicate an impairment in labial ROM. If the lips cannot close symmetrically, an impairment in labial range of motion is indicated.

2. *Labial strength.*

- Ask the patient to pucker his or her lips.

- Provide resistance with a tongue depressor against the patient's lips while they are puckered.

- Ask the patient to press the lips around the tongue depressor and hold it.

- Try to pull the tongue depressor out.

- Note whether one side has greater labial strength than the opposite side.

☛ If the patient is unable to pucker his or her lips symmetrically, impairment in labial strength is indicated. If the patient cannot maintain puckered lips with resistance from the tongue depressor, impairment in labial strength is indicated. If the patient cannot hold the tongue depressor with the lips as the therapist attempts to pull it away, impairment in labial strength is indicated.

3. *Lingual ROM and strength.*

- Ask the patient to protrude his or her tongue.

- Apply resistance to the protruded tongue with a tongue depressor.

- Ask the patient to retract the tongue.

- Apply resistance to the retracted tongue with a tongue depressor.

- Ask the patient to elevate the tongue.

- Apply resistance to the elevated tongue with a tongue depressor.

- Ask the patient to move the tongue laterally (right and left).

- Apply resistance to the laterally protruded tongue with a tongue depressor.

- Note whether one side has greater ROM and strength than the opposite side.

☞ If the patient has difficulty protruding, retracting, elevating, and laterally moving his or her tongue, impairment in lingual ROM is indicated. If the patient cannot perform these movements symmetrically, impairment is indicated. If the patient has difficulty resisting these tongue movements, impairment in lingual strength is indicated.

4. *Mandibular ROM and strength.*

- Ask the patient to open and close his or her mouth.

- Apply resistance to the mandible while asking the patient to keep the mouth open.

- Note whether one side of the face has greater ROM and strength than the opposite side.

☞ If asymmetry is noted when the patient opens and closes his or her mouth, impairment in mandibular ROM is indicated. If one side of the mandible has greater strength than the other, impairment is indicated.

5. *Observation of the soft palate.*

- Observe the soft palate at rest, upon mouth opening. Note symmetry.

- Ask the patient to say "a." Note symmetry of soft palate elevation.

☞ If the velum is asymmetrical at rest, impairment is indicated. If soft palate elevation is asymmetrical, impairment is indicated. Limited to no velar movement may lead to nasal regurgitation.

6. Gag reflex.

■ Stimulate the faucial arches with a tongue depressor to elicit the gag reflex. The gag reflex often presents as head and jaw extension, tongue protrusion, and pharyngeal contraction.

☞ An absent or sluggish gag reflex may indicate impairment in swallowing function; however, recent literature suggests that the gag reflex may not be important for normal swallowing to occur.

Laryngeal Examination

The major task of the larynx in swallowing is to guard against entrance of any material in the airway that would be harmful to the lungs. This task is accomplished by laryngeal elevation. Laryngeal examination involves palpating for laryngeal elevation, listening to a patient's vocal quality, and asking the patient to cough.

1. Laryngeal elevation.

■ Place two fingers (the second and third phalanges) lightly on the patient's throat and feel for laryngeal movement as the patient swallows.

■ Place the fingers perpendicular to the patient's neck with the index finger on top of the thyroid (see Figure 9.1). The patient's larynx should elevate above the index finger on swallowing. Note that this is a subjective measure of laryngeal elevation.

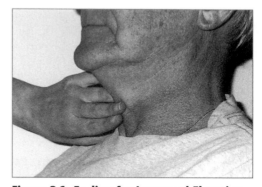

Figure 9.1 Feeling for Laryngeal Elevation

☞ If the patient's larynx does not elevate above the therapist's index finger on swallowing, dysphagia may be indicated.

2. Vocal quality. Listening to a patient's voice quality both before and after food trials is of paramount importance because one of the best indictors of a swallowing disorder is a wet, gurgly voice.

■ Ask the patient to repeat the following phrases: "In 1492 Columbus sailed the ocean blue." "A penny saved is a penny earned." If the patient has difficulty with following directions or has aphasia, be sure to observe the patient's vocal quality when he or she verbalizes independently.

☞ A wet voice that is present before the administration of food or liquid indicates poor management of oropharyngeal secretions. The dysphagia screening should be terminated at this point because poor vocal quality is a sure sign of dysphagia.

3. *Coughing.*

■ Ask the patient to cough. The ability to cough demonstrates the larynx's ability to close.

☞ A weak cough usually results from reduced vocal fold or laryngeal closure and indicates dysphagia.

Observation With Food Trials

The administration of *varied food textures* should be done only after a detailed oral–motor examination is performed. Food trials should *not* be given to a patient who is lethargic, cannot be positioned properly, or has a wet vocal quality. Signs of aspiration and/or penetration should be monitored both immediately after each swallow and a couple of minutes after each swallow.

Therapists usually have differing opinions or rationales regarding what type of food consistency should be used to begin testing at bedside. If a patient has significant labial and lingual weaknesses it is advisable to begin with a *puree consistency* because liquid may be difficult to control. Avoid giving *hard solids* to a patient who is edentulous (lacking teeth) or whose dentures are loose because the patient is at a higher risk of choking. *Thin liquids* may be given at the end of the evaluation once other consistencies have been tested. The rationale is that thin liquids are usually the most problematic for patients with dysphagia, and initiating an evaluation with them may cause the patient difficulty early in the screening. *Thick liquid* trials should be attempted if the patient displays difficulty with thin liquids.

RED FLAGS
SIGNS AND SYMPTOMS OF DYSPHAGIA

A red flag is an indicator or symptom of dysfunction. If a patient is observed displaying any of the following red flags, it is likely that the patient possesses dysphagia.

ORAL PREPARATORY STAGE

■ Cannot remove or strip food from the utensil

■ Food falling out of oral cavity

■ Resistance to eating or drinking

■ Rapid rate of food and liquid intake

■ Vocalizing with food in the oral cavity.

ORAL STAGE

- Drooling

- Residue on the tongue

- Pocketing or collecting food in the side of the mouth.

PHARYNGEAL STAGE

- Coughing

- Choking

- Throat clearing

- Wet vocal quality

- Increased respiratory rate

- Reports of pain on swallowing

- Nasal regurgitation.

INSTRUMENTAL ASSESSMENTS FOR DYSPHAGIA

When a dysphagia screening indicates the need for further instrumental or objective assessment, the most widely used assessments include the modified barium swallow/videofluoroscopy and fiberoptic endoscopic examination of swallowing with sensory (FEES) testing.

MODIFIED BARIUM SWALLOW/VIDEOFLUOROSCOPY

This test is designed to study the anatomy and physiology of all stages of swallowing. It provides information regarding the cause of the swallowing dysfunction and aids in identifying which strategies and dietary modifications can improve swallowing function. A speech–language pathologist or an occupational therapist in conjunction with a radiologist usually conducts this test. The following procedure is used:

- The patient is seated upright, and a lateral view is taken. An anterior–posterior view is used when looking at the symmetry of the swallow, esophageal transit, and gross amounts of reflux.

- Varied food consistencies (thin liquid, thick liquid, puree, soft and hard solids) are impregnated with barium for contrast and administered in calibrated amounts.

- Images are recorded on videotape; a microphone is used to record dialogue and to differentiate between specific consistencies given and compensatory strategies used.

The advantages of this test are that it

- Permits viewing of all stages of the swallow

- Allows viewing of the anatomy in different planes (lateral and anterior–posterior)

- Can be used to measure the effects of compensatory strategies.

The disadvantages of this test are that it

- Exposes the patient to radiation

- Is sometimes difficult to transport medically fragile patients to a radiology suite.

FEES TESTING

This test involves passing a fiberoptic laryngoscope transnasally to the hypopharynx, where the larynx and surrounding structures can be viewed. A speech–language pathologist extensively trained to do flexible fiberoptic laryngoscopy and with expertise in the evaluation of swallowing usually performs this test. The equipment for the test includes the flexible fiberoptic endoscope, camera, appropriate light, source recorder, monitor, and microphone. For a detailed description of FEES equipment and other swallowing assessments, refer to *Dysphagia: A Continuum of Care* (Sonies, 1997). The following procedure is used:

- The fiberoptic scope is inserted transnasally, and the patient is given various bolus volumes and viscosities.

- Food consistencies are usually colored with food dye for contrast.

- The events that occur before and after the swallow are observed.

- A pulse of air is delivered through the laryngoscope to examine sensory levels in the pharynx. The air is delivered to the side of each arytenoid in order to elicit a laryngeal adductor reflex. If the reflex is absent, decreased sensation is suspected, and the patient is considered to be at a higher risk of silent aspiration.

The advantages of this test include the following:

- No travel stress on the patient because the FEES is portable

- No exposure to radiation

- Allows viewing of oral secretions.

The disadvantages of this test include the following:

- The oral stage of the swallow cannot be observed.

- Movement of the structures within the hypopharynx prevent viewing of events during the swallow.

- The procedure is invasive.

- The patient must be somewhat cooperative. For example, it is difficult to perform this test on patients with movement disorders or who are combative.

AVAILABLE DYSPHAGIA SCREENINGS

Burke Dysphagia Screening Test (BDST)
J. R. Odderson, B. A. McKenna
Burke Rehabilitation Hospital (1993)
785 Mamaroneck Avenue
White Plains, NY 10605
(914) 597-2500

RIC Clinical Evaluation of Dysphagia (CED)
L. Cherney, C. Cantieri, J. Pannell
Rehabilitation Institute of Chicago (1986)
345 East Superior Street
Chicago, IL 60611
(312) 238-1000

Reference
Sonies, B. (1997). *Dysphagia: A continuum of care.* Gaithersburg, MD: Aspen.

DYSPHAGIA SCREENING

Patient Name _____   Date _____

Diagnosis _____

Current Diet _____

ORAL PERIPHERAL EXAMINATION

Facial tone	☐ Intact	☐ Impaired	Comments
Labial ROM	☐ Intact	☐ Impaired	Comments
Labial strength	☐ Intact	☐ Impaired	Comments
Lingual ROM	☐ Intact	☐ Impaired	Comments
Lingual strength	☐ Intact	☐ Impaired	Comments
Mandibular ROM	☐ Intact	☐ Impaired	Comments
Mandibular strength	☐ Intact	☐ Impaired	Comments
Soft palate symmetry	☐ Intact	☐ Impaired	Comments
Soft palate movement	☐ Intact	☐ Impaired	Comments
Soft palate dentition	☐ Intact	☐ Impaired	Comments
Gag reflex	☐ Intact	☐ Impaired	Comments
Oral secretions	☐ Intact	☐ Impaired	Comments

OBSERVATION

Consistencies tested			
Preoral/oral stage			
Acceptance	☐ Intact	☐ Impaired	Comments
Rate of intake	☐ Intact	☐ Impaired	Comments
Anterior containment	☐ Intact	☐ Impaired	Comments
Bolus formation	☐ Intact	☐ Impaired	Comments
Oral transit	☐ Intact	☐ Impaired	Comments
Pharyngeal stage			
Swallow triggering	☐ Intact	☐ Impaired	Comments
Laryngeal rise	☐ Intact	☐ Impaired	Comments
Coughing	☐ Yes	☐ No	Comments
Throat clearing	☐ Yes	☐ No	Comments
Wet, gurgly voice	☐ Yes	☐ No	Comments
Other signs of aspiration	☐ Yes	☐ No	Comments

ASSESSMENT

Swallowing function		
Oral preparatory stage	☐ Intact	☐ Impaired
Oral stage	☐ Intact	☐ Impaired
Pharyngeal stage	☐ Intact	☐ Impaired
Comments		

_____
_____
_____
_____

RECOMMENDATIONS

☐ Instrumental assessment

☐ Speech/swallowing clinical bedside evaluation

☐ No further intervention indicated

Glossary

Abulia: A patient lacks the impulse to move, although is not paralyzed. *See* catatonia.

Acalculia: The inability to calculate mathematical equations. *See also* dyscalculia.

Accommodation: A three-step process involving a change in the thickness of the lens, convergence of the eyes, and pupillary constriction. Accommodation enables one to adjust to changes in *focal length* (quickly changing focus from near to far objects) and achieve the sharpest focus.

Active range of motion (AROM): The amount of joint motion achieved during unassisted voluntary joint motion.

Acuity: The ability to see visual detail; the accuracy of sight.

Agraphesthesia: The inability to interpret letters written on the palmar surface of one's hand.

Agraphia: The inability to write intelligible words and sentences; the written form of *alexia* (the inability to read).

Agrommation: The inability to arrange words sequentially so that they form intelligible sentences in conversation or writing.

Ahylognosia: The inability to discriminate between different types of materials by touch alone.

Akathisia: An inability to remain still, caused by an intense urge to move or fidget.

Akinesia: An inability to initiate voluntary movement; it is commonly seen in the late stages of Parkinson's disease. Generally, the patient assumes a fixed posture as a result of an inability to initiate movement. Patients report that a tremendous amount of mental concentration is required to perform the most basic motor tasks.

Alexia: The inability to comprehend the written word; an inability to read. *See also* dyslexia.

Alexithymia: A type of expressive aphasia; an inability to attach words to one's emotions.

Amorphagnosia: The inability to discriminate between different forms by touch alone.

Anomia: The inability to remember and express the names of people and objects. The patient may know the person but cannot remember his or her name. *See also* prosopagnosia.

Anosmia: Loss of smell.

Anosognosia: An extensive neglect syndrome involving failure to recognize one's paralyzed limbs as one's own. It results from lesions of the right cerebral hemisphere.

Anterograde amnesia: A decreased ability to remember ongoing day-to-day experiences after the injury has occurred. It involves a decreased ability to encode short-term memory into long-term storage.

Ape (simian) hand: Occurs as a result of denervation and its subsequent atrophy of muscles of the thenar eminence. Thumb opposition is lost. The hand flattens as a result of atrophy and paralysis. Occurs as a result of median nerve damage.

Aphasia: An impairment in the expression and/or the comprehension of language. *See* expressive aphasia and receptive aphasia.

Apraxia: Motor planning impairment. *See* ideational apraxia and ideomotor apraxia.

Aprosodia: Impaired comprehension of tonal inflections used in conversation. Patients have difficulty perceiving the emotional tone of someone's verbal expression. Aprosodia often results from a right cerebral hemisphere lesion. Patients can often understand the literal meaning of words but cannot interpret the words' emotional tonal inflections.

Associated reactions: Stereotyped reactions in which the effortful use of one extremity influences the posture and tone of the opposite (or another) extremity; unintentional movements of one extremity during the voluntary movement of another extremity. For example, a patient several weeks poststroke, with a right upper extremity in a spastic flexor synergy, uses the uninvolved left upper extremity to brush his or her hair. Simultaneously, the involved right upper extremity becomes more flexed and adducted. Associated reactions often occur during voluntary strenuous or effortful movement. For example, it is common to observe an upper-extremity spastic flexor pattern become more hypertonic when the patient attempts to ambulate (effortful movement). Associated reactions can occur in both the involved and uninvolved extremities.

Astereognosis: The inability to identify objects by touch alone.

Asthenia: Muscle weakness; a sign of cerebellar damage.

Asymbolia: Difficulty comprehending gestures and symbols.

Ataxia: An umbrella term used to describe incoordinated patterns of movement that affect one's gait, posture, and upper-extremity motor control. An ataxic gait is characterized by a wide base of support with arms held away from the body to enhance balance. Ambulation is unsteady, and the patient appears to stagger as he or she walks. The patient also is unable to maintain a straight, direct forward line while walking and instead tends to veer toward the side of the lesion. An ataxic posture is characterized by back-and-forth oscillations of the body while standing upright. Ataxic patterns of the upper extremities appear as up-and-down oscillating movements when the limbs are held against gravity.

Athetosis: Or *athetoid movements,* are slow–flailing, twisting movements that are wormlike in quality. Athetosis often presents in combination with spasticity or hypertonicity and is believed to result from damage to the caudate and/or putamen. Athetoid movements commonly involve the neck, face, trunk, and extremities. Athetosis is a clinical feature of cerebral palsy.

Audition: The sensation of hearing, mediated by the vestibulocochlear nerve (cranial nerve 8) and the primary auditory cortex.

Benediction sign: Occurs as a result of paralysis to the flexors of the third and fourth digits. When the patient is asked to make a fist, the third and fourth digits do not fully flex; instead, these fingers remain in a position similar to the benediction sign made by clergy. Results from median nerve damage.

Bicycle rider's syndrome: Prolonged bicycle riding can cause compression of the ulnar nerve in the hypothenar eminence.

Binocular fusion: The use of both eyes together to produce a fused single image.

Blepharospasm: A torsion dystonia in which the eyes are involuntarily kept closed.

Body schema: The awareness of spatial characteristics of one's own body in space, the relationship of individual body parts to each other, and the relationship of the body to the environment. Body schema is derived from the synthesis of tactile, proprioceptive, and pressure sensory perceptions.

Bradykinesia: Slowed or decreased movement; it is sometimes referred to as *poverty of movement.* In addition to slowed movements, the patient's ability to quickly change movements becomes delayed—switching from one motor pattern to another becomes difficult. Bradykinesia also can manifest as a lack of facial expression (or masked face), monotone speech, and reduced eye movement.

Bradyphrenia: Slowness of thought; commonly seen in Parkinson's disease (in late stages), schizophrenia, and depression. Also referred to as *poverty of thought. Aphrenia* is a complete stoppage of thought.

Brain death: A type of coma that occurs when all brainstem functions are lost; coma characterized by sleeplike (eyes-closed) unarousability due to extensive damage to the reticular activating system. All vegetative functions are lost, leading to fatal respiratory infections.

Broca's aphasia: An expressive language disorder in which patients can understand what is spoken to them, but they cannot express their ideas in an understandable way. Broca's aphasia always results from a left cerebral hemisphere lesion in the brain region referred to as *Broca's area.*

Carpal tunnel syndrome: The most common median nerve entrapment syndrome. The median nerve is compressed as it travels through the tunnel formed by the two rows of carpal bones and the *flexor retinaculum* (the fibrous ligamentous band that forms the carpal tunnel). In women, carpal tunnel syndrome may result from the hormonal changes occurring during pregnancy. Overuse syndromes and rheumatoid arthritis may also cause compression of the median nerve at the carpal tunnel.

Catatonia: Lack of motivation to move, although the patient is not paralyzed. *See also* abulia.

Causalgia: An intense burning pain accompanied by trophic skin changes—on the region of denervation the skin becomes smooth and glossy, the nails become thick, and hair thins or falls out.

Chorea: Or *choreiform movements,* are sudden, rapid, involuntary jerky movements that primarily involve the face and extremities. Chorea is associated with Huntington's disease—a degenerative disorder involving the caudate and putamen. Shoulder shrugs, hip movements, crossing and uncrossing one's legs, facial grimaces, and tongue protrusions are signs of chorea.

Clasp knife phenomenon: A form of hypertonicity in which the joint, at first, cannot be moved when passively ranged. A sustained stretch causes the hypertonicity to suddenly give way, allowing the joint to be passively ranged.

Clawhand deformity: Occurs as a result of ulnar nerve damage. The first phalanx is hyperextended, and the distal two phalanges are flexed. The fifth digit is abducted. There is an inability to voluntarily abduct or adduct the fingers. Weakened thumb opposition and total loss of fifth finger opposition occur as well. Atrophy of the intrinsic muscles of the fingers and flattening of the hypothenar eminence also occur.

Clonus: Uncontrolled oscillation of a spastic muscle group that results from a quick muscle stretch.

Cognition: The ability to use mental processing skills to interact with and meet the demands of one's environment. Mental processing skills extend on a continuum from low-level skills (e.g., alertness/arousal, orientation, recognition, initiation/termination of activity) to more complex, high-level skills (e.g., sequencing multiple steps, categorizing multiple items, problem solving, planning, generalization of new learning, and insight).

Color agnosia: An inability to remember the appropriate colors for specific objects. Patients with color agnosia appear to forget the color of common objects. For example, a patient with color agnosia may believe that a banana is blue.

Color anomia: An inability to remember the names of colors. Color anomia differs from color agnosia in that patients with color anomia may forget the names for colors, but they would still recognize that a banana is not blue.

Coma: A state in which a patient experiences a loss of consciousness characterized by a loss of awareness of self and environment and an inability to respond to external stimuli or internal drives. There are differing degrees of coma: profound coma, semi-coma, stupor, minimally conscious, drowsy–confused, persistent vegetative state, and brain death.

Common peroneal nerve damage: Lesions to the common peroneal nerve can result in an inability to dorsiflex or evert the foot. To compensate for a lack of dorsiflexion, patients use excessive hip flexion, resulting in a steppage gait. Foot drop commonly is observed. Sensation often is lost or diminished on the lateral portion of the leg and on the dorsum of the foot. Pain rarely is felt. Lesions to the common peroneal nerve can be differentiated from sciatic nerve lesions: With common peroneal nerve damage, the Achilles tendon reflex is preserved; with sciatic nerve lesions, the Achilles tendon reflex is absent.

Confabulation: The generation of false information to account for memories that the patient is unable to recall. Patients with brain damage commonly generate intricate and complex false stories in an attempt to fill in missing parts of their memory.

Consensual light reflex: Also referred to as *consensuality,* occurs when the pupil of the eye that is not receiving light constricts at the same time as the pupil of the eye that is receiving light. This reflex occurs because the optic nerve of the eye that is being stimulated carries the light stimulus message to the brain. Both oculomotor nerves are then stimulated, resulting in constriction of the pupils in both eyes.

Contractures: Limitation in joint movement due to shortening of muscles, tendons, and ligaments; results from inactivity at a joint. Often occurs secondary to spasticity and rigidity.

Contralateral homonymous hemianopsia: A bilateral field deficit in the temporal or nasal field of both eyes due to an optic tract lesion; a loss of the visual field on the opposite side of the lesion. Also referred to as a *field cut.*

Contrast sensitivity: The ability to visually detect subtle changes in contrast between the background and foreground.

Convergence: The ability of both eyes to move medially during a near task, allowing a patient to perform such tasks as reading.

Convergence insufficiency: An inability of both eyes to move medially during a near task; can be indicated by complaints of losing one's place when reading or writing and difficulty performing tasks close up.

Corneal reflex: A blink reflex elicited when some substance wipes against the cornea of the eye.

Cubital tunnel syndrome: The most common entrapment syndrome at the elbow that results from ulnar nerve damage. Elbow deformities, repetitive motion at the elbow, inflammatory conditions, and trauma to the elbow may all cause cubital tunnel entrapment.

Cunctation-festinating gait: *Cunctation* means to resist movement. *Festination* means to hurry. A cunctation-festinating gait is characterized by difficulty both initiating and stopping walking. Once the patient is able to begin walking, the movement patterns for walking become hurried. The patient appears to quickly shuffle. Reciprocal arm swing often is absent; however, sometimes a patient may demonstrate an exaggerated arm swing to enhance propulsion of movement. The patient also demonstrates an inability to stop walking once started and often bumps into walls or furniture. Changing directions while walking is a difficult task—patients often are unable to circumvent obstacles in their path once they have initiated a particular walking direction. Balance and equilibrium responses often are decreased or absent. A cuncta-tion-festinating gait commonly is seen in Parkinson's disease.

Deep tendon reflex: A reflex that is mediated by the reflex, or spinal, arc (at the spinal cord level) and does not require feedforward or feedback from the cortex. Deep tendon reflexes include the following reflexes: biceps tendon reflex, brachioradialis tendon reflex, triceps tendon reflex, patellar tendon reflex, Achilles tendon reflex, and the plantar reflex.

Dementia: A progressive, irreversible, cognitive deterioration. Signs include memory deficits (particularly short-term memory deficits in the early stages of dementia and long-term memory deficits in the end stages), emotional lability, changes in personality, and lack of insight into cognitive deficits.

Depth perception dysfunction: Or *stereopsis*, is an inability to determine whether objects in the environment are near or far in relation to each other and in relation to the patient.

Dermatome: A skin segment innervated by a specific peripheral spinal nerve.

Diplopia: Double vision—either horizontal or vertical. Diplopia also can refer to blurred or shadowed vision.

Disconjugate gaze: An inability to move both eyes in equal relation to one another. Usually results in a form of diplopia or focusing difficulties.

Disinhibition: The loss of the ability to self-regulate socially inappropriate behaviors; it often occurs as a result of orbitofrontal lobe damage. The patient may display impulsivity, aggression, irritability, agitation, and/or sexual disinhibition.

Disorientation: Or *confusion*, is demonstrated by a patient's lack of understanding of present events and disorientation regarding who the patient is, where the patient is, why the patient is in the hospital, and what month or season it is. Patient orientation is assessed with regard to person, place, and time.

Dressing apraxia: Involves an inability to dress oneself due to either a body schema disorder or an apraxia.

Dysarthria: Difficulty enunciating words; slurred speech, usually due to an impairment of the cranial nerves innervating the muscles of speech. Often due to cerebellar damage.

Dyscalculia: Difficulty calculating mathematical equations. *See also* acalculia.

Dysdiadochokinesia: An impaired ability to perform rapid alternating movements, such as bilateral forearm pronation–supination or bilateral hand grasp–release. The patient's attempt at rapid alternating movements becomes irregular; bilateral movements cease to be simultaneous. Often, one extremity lags behind the other. A sign of cerebellar damage. *Adiadochokinesia* refers to the absence of the ability to perform rapid alternating movements.

Dysesthesia: The occurrence of unpleasant sensations, such as burning.

Dysgraphia: Difficulty writing because the patient cannot break down words into their most basic units—phonemes. Dysgraphia is the written form of *dyslexia* (difficulty reading because the patient cannot break down words into phonemes).

Dyslexia: The impaired ability to read. Dyslexia is a language problem in which the ability to break down words into their most basic units—phonemes—is impaired. The patient may perceive letters as reversed or sequentially mixed up. Some words in a sentence may be overlooked or left out. *See also* alexia.

Dysmetria: An inability to judge the distance and range of a movement. It is characterized by overshooting (past pointing) or undershooting one's reach for a target object. A sign of cerebellar damage.

Dysphagia: Difficulty swallowing, leading to choking and food aspiration (occurs when food travels down the trachea instead of the esophagus tube).

Dysphonia: Decreased phonal volume or hoarseness of voice. *Aphonia* is an inability to make sounds. *Hypophonia* refers to reduced vocal force.

Dyspnea: Breathing difficulties. *Apnea* is arrested breathing.

Dyssynergia: Also referred to as *movement decomposition.* Dyssynergia is characterized by movements that are broken up into their component parts rather than as a smooth, single movement. A sign of cerebellar damage.

Dystonia: A movement disorder resulting from increased muscle tone causing distorted, twisted postures of the trunk and proximal extremities. The sustained muscle contractions of dystonia—referred to as *torsion spasms*—can last from seconds to hours. Common torsion spasms include blepharospasm, torticollis, and truncal dystonia (see definitions of individual terms).

Esotropia: Internal or medial strabismus due to a lesion of cranial nerve 6 (abducens nerve).

Exotropia: External or lateral strabismus due to a lesion of cranial nerve 3 (oculomotor nerve).

Expressive aphasia: The inability to express language in a clear, fluent, meaningful way.

Extinction of simultaneous stimulation: The inability to determine that one has been touched on both involved and uninvolved sides—the neural sensation of the uninvolved side overrides the ability to perceive touch on the involved side.

Extraocular motility: Is comprised of six eye muscles (four rectus and two oblique) that allow for eye movement (or motility) within the six cardinal planes.

Extraocular range of motion: The degree of range of motion present in each eye within all six cardinal planes—the vertical and horizontal (or temporal) planes, and the four quadrants (superior, inferior, right, and left planes).

Figure–ground discrimination dysfunction: An inability to distinguish objects in the foreground from objects in the background.

Finger agnosia: An impaired perception of the relationship of the fingers to each other.

Flaccidity: Loss of muscle tone due to denervation of specific peripheral nerves. Also occurs in muscles innervated at the level at which the spinal cord was severed in a spinal cord injury.

Foot drop: Results from denervation of the dorsiflexors. The patient often drags a foot along the ground during ambulation.

Form–constancy dysfunction: An inability to recognize subtle variations in form or changes in form, such as a size variation of the same object.

Functional muscle strength: The amount of resistance a joint can sustain during a movement.

Gag reflex: Causes choking and regurgitation when the soft palate, base of the tongue, palatine arches, glossopalatine arch, and/or pharyngopalatine arch are stimulated. Often presents as head and jaw extension, tongue protrusion, and pharyngeal contraction.

Generalization of learning: The ability to transfer the skills needed for one task to a new task that is similar.

Grip strength: The ability to maintain a hand grasp over a period of time.

Gustation: The sensation of taste, mediated by the facial nerve (cranial nerve 7) and the glossopharyngeal nerve (cranial nerve 9).

Guyon's canal (ulnar tunnel) syndrome: A condition in which the ulnar nerve becomes entrapped as it travels from the forearm into the hand through Guyon's canal (the ulnar tunnel). The floor of the tunnel is formed by the pisiform and hamate bones; the upper portion of the tunnel is formed by the pisohamate ligament. Overuse of or damage to the hypothenar eminence of the hand often results in compression of Guyon's canal.

Hemianopsia: Or field cut. *See* contralateral homonymous hemianopsia.

Hemiballismus: Characterized by violent thrashing movements of the extremities on one side of the body (the side that is contralateral [opposite] to the lesioned basal ganglia). It results from a lesion to the subthalamus and caudate.

Hemiparesis: Partial paralysis or muscular weakness of limbs on one side of the body. Occurs on the contralateral side (opposite side) of the lesion site.

Hemiparesthesia: Loss of sensation of limbs on one side of the body. Occurs on the contralateral side (opposite side) of the lesion site.

Hemiplegia: Complete paralysis of limbs on one side of the body. Occurs on the contralateral side (opposite side) of the lesion site.

Hyperesthesia: An increase in the perception of sensation.

Hyperkinesia: Disorders involving speeded movement (e.g., chorea, athetosis). Opposite of akinesia.

Hyperopia: Farsightedness. Opposite of nearsightedness, or myopia.

Hyperreflexia: An increase in deep tendon reflexes.

Hypertonicity: An increase in muscle tone.

Hypertropia: Occurs when one eye is deviated superiorly on forward gaze; in other words, one eye deviates upward compared with the other eye. Corneal reflection falls on the inferior rim of the pupil.

Hypoesthesia: An absence or decrease in the perception of sensation.

Hypokinesia: The slowing of movement just short of complete loss of movement (or akinesia).

Hyporeflexia: A decrease in or absence of deep tendon reflexes.

Hypotonicity: A decrease in muscle tone.

Hypotropia: Occurs when the eye is deviated inferiorly on forward gaze; in other words, one eye deviates downward compared with the other eye. Corneal reflection falls on superior rim of the pupil.

Ideational apraxia: Motor planning impairment involving an inability to cognitively understand the motor demands of the task.

Ideomotor apraxia: Involves the loss of the kinesthetic memory of motor patterns; in other words, the motor plan for a specific task may be lost. Or, the motor plan may be intact, but the patient cannot access the appropriate motor plan and may implement an inappropriate motor plan for a specific task.

Intention tremors: Occur during voluntary movement of a limb and tend to increase as the limb nears its intended goal. For example, a patient may experience increased tremors as he or she attempts to use the hand to bring a spoon to the mouth. *Tremor* is the rhythmic oscillation of joints caused by alternating contractions of opposing muscle groups. Intention tremors tend to diminish or stop when the patient's limbs are at rest. A sign of cerebellar damage.

Kinesthesia: The ability to identify one's limbs as they move through space, with vision occluded.

Locked-in syndrome: Occurs when a comatose patient has full receptivity to external stimuli but lacks the ability to make any response. Most often it is due to a lesion in the basil pons, which interrupts the descending motor pathways but spares the ascending sensory pathways.

Long-term memory or remote memory: The ability to retrieve stored memories of events and information that occurred more than 48 hours ago. Long-term memory often is spared or returns as a patient recovers from brain injury.

Masseter reflex: A contraction that occurs when the masseter muscle is lightly tapped with a reflex hammer.

Metamorphopsia: A visual distortion of the physical properties of objects so that objects appear bigger, smaller, heavier, or lighter that they really are.

Micrographia: Microscopic handwriting; often seen in patients with Parkinson's disease.

Mononeuropathy: A form of neuropathy involving damage to a single peripheral nerve.

Mononeuropathy multiplex: A form of neuropathy involving damage of multiple single peripheral nerves in an asymmetric random pattern. Common causes include diabetes mellitus and polyarteritis.

Motor impersistence: Occurs when a patient attempts to maintain both extremities in the same position, but unbeknownst to the patient, the involved extremity drifts out of its position. A sign of cerebellar damage.

Movement decomposition. *See* dyssynergia.

Muscular tone: Resistance of a muscle to stretch.

Myopia: Nearsightedness—the inability to see detail at a distance. Opposite of farsightedness, or hyperopia.

Myotome: A muscle innervated by a specific peripheral nerve; closely follows the pattern of the dermatomes.

Neuropathy: Involves some form of impairment occurring within the peripheral nervous system. There are five major categories of peripheral neuropathy: polyneuropathy, mononeuropathy, mononeuropathy multiplex, plexopathy, and radiculopathy (see individual definitions).

Nystagmus: A condition in which the eyeballs involuntarily move back and forth in a quick, jerky, oscillating fashion when the eyes move (a) laterally or medially to either the temporal of nasal field extremes or (b) superiorly and inferiorly in the visual field extremes.

Object fixation: The ability to locate and focus on a stationary object or target.

Olfaction: The sensation of smell; mediated by the olfactory nerve (cranial nerve 1).

Optokinetic reflex: A visual reflex that is stimulated by visual movement. A positive optokinetic reflex produces nystagmus, indicating some level of vision.

Paresis: Partial paralysis.

Paresthesia: The occurrence of unusual sensations, such as "pins and needles."

Passive range of motion: The amount of available joint motion when the therapist moves the joint through full range of motion.

Perception: The ability to interpret or attach meaning to sensory information from the external and internal environments. Perceptual impairment more often involves dysfunction of the right cerebral hemisphere than the left; although some perceptual disorders result from left hemisphere damage (e.g., some perceptual language disorders). Right hemisphere perceptual disorders involve a distortion of the physical environment—the patient's perception of the environment and his or her body become distorted. There are several classifications of perception: visual perception, visual–spatial perception, tactile perception, auditory perception, body schema perception, language perception, and motor perception.

Perseveration: The inability to stop an activity once started. Patients are unable to interpret cues indicating that they need to stop the task or change strategies. Instead, they continue to implement the behavior over and over.

Persistent vegetative state: A type of coma characterized by eyes-open unconsciousness. The patient is able to blink to threat and is capable of a few primitive postural movements but is otherwise without awareness and unresponsive to external stimuli. In persistent vegetative states, the brainstem and its vegetative functions (i.e., cough, gag, and swallowing reflexes) are spared. The brunt of neurologic damage is to the cerebral hemispheres. Thus, life expectancy is lengthened—the patient can remain alive in this state indefinitely with technological life support.

Pinch strength: The ability to maintain a fingertip grasp over a period of time.

Piriformis syndrome: Occurs when the sciatic nerve is compressed between the piriformis and the obturator internus muscles as the nerve exits the pelvis in the greater sciatic foramen.

Plexopathy: A form of neuropathy involving damage to the peripheral nerves in one of the plexi—brachial, lumbar, or sacral plexus. Idiopathic brachial neuritis and traumatic injury to the brachial plexus are common causes.

Polyneuropathy: A form of neuropathy involving bilateral, symmetric damage to the peripheral nerves. Usually the lower extremities are more commonly involved than the upper extremities. The distal segments of an extremity are involved first, before the proximal segments. Signs and symptoms commonly start in the feet and move proximally to the knees. Often, symptoms are present at the knee level before they are found in the hands. The typical pattern of presentation is stocking-and-glove neuropathy in which the feet and knees are affected first, followed by the hands.

Position in space dysfunction: Involves difficulty using concepts relating to positions, such as up/down, in/out, behind/in front of, and before/after.

Primitive reflexes: Reflexes that humans are born with or that develop and become integrated by the central nervous system in infancy and toddlerhood. These reactions facilitate gross motor patterns of flexion and extension. Adults who exhibit primitive reactions usually have sustained severe brain damage. Primitive reflexes are mediated by specific areas (or levels) of the central nervous system: spinal cord, brainstem, basal ganglial, and cortical.

Proprioception: The ability to identify one's trunk and limb position in space when vision is occluded.

Prosopagnosia: The inability to identify familiar faces because the patient cannot perceive the unique expressions of facial muscles that make each human face different from another.

Ptosis: Drooping of the upper eyelid, as in Horner's syndrome (due to a lesion of cranial nerve 3).

Pupillary constriction: Or *pupillary reflex,* occurs when light is shined into an eye, causing the pupil to constrict.

Pupillary size: The aperture that controls the amount of light entering the eye; measured in millimeters.

Radiculopathy: A form of neuropathy involving spinal nerve root damage to the dorsal or ventral roots. Herniated vertebral discs and vertebral bone disease are common causes.

Rebound phenomenon: The inability to regulate the action of opposing muscle groups. The patient is asked to resist the therapist's attempt to pull the patient's flexed elbow into extension. The therapist then releases the patient's forearm. Normally, the elbow should remain in approximately the same position due to the regulation of opposing muscle groups. Patients with cerebellar lesions are unable to regulate their opposing muscle groups, and their limb will suddenly hit their torso. This occurs as a result of impaired proprioceptive feedback—the patient cannot regulate the speed and force of opposing muscle groups quickly enough to prevent the arm from hitting the torso. A sign of cerebellar damage.

Receptive aphasia: The impairment in the comprehension of language.

Reflex sympathetic dystrophy: Or *frozen shoulder,* is characterized by immobility and pain in an upper extremity having nerve damage.

Resting tremors: Or *nonintention tremors,* are involuntary oscillating movements that occur in an extremity at rest. Resting tremors decrease or disappear with the initiation of voluntary movement and tend to worsen with increased emotional stress. Often associated with Parkinson's disease (a basal ganglia disorder) and characterized by a *pill-rolling movement*—a tremor in which the patient appears to be rolling a pill between the thumb and first two fingers. Resting tremors also may be observed at the wrist in the form of hand tremors and at the forearm in the form of pronation–supination tremors. A head tremor also may be observed.

Retrograde amnesia: The loss of one's entire personal past after traumatic injury to the brain. It is a common consequence of brain injury. Long-term memory often returns to patients as they recover from brain injury.

Right–left discrimination dysfunction: An inability to use the concepts of right and left accurately.

Rigidity: A form of *hypertonicity* (increased muscle tone) that is characterized by increased resistance to passive movement of a joint in all planes. Rigidity differs from *spasticity* in that spasticity involves increased tone in either the flexors (abductors) or the extensors (adductors) of a joint, but not both. Two types of rigidity are signs of basal ganglia impairment: (a) *lead pipe rigidity* is characterized by a uniform and continuous resistance to passive movement as the extremity is moved through its range of motion (in all planes) and (b) *cogwheel rigidity* is characterized by an alternate release/resistance pattern to passive movement as the extremity is moved through its range of motion (in all planes). Cogwheel rigidity can be felt as a series of brief muscle relaxations followed by quick contractions and is a common sign of Parkinson's disease.

Saccades: Quick, precise eye movements that are made during visual scanning or a visual search.

Saturday night palsy: Results from compression of the radial nerve at the spiral groove. The name is derived from the experience of an individual—who is inebriated—falling asleep with an arm against a rigid surface (e.g., a park bench or sidewalk), causing all muscles innervated by the radial nerve to become denervated (except the triceps muscle).

Sciatic nerve damage: Results in a loss of voluntary knee flexion secondary to denervation of the hamstring muscles. Diminished innervation of all the muscles of the leg and foot results in a steppage gait (the compensation for a lack of dorsiflexion by using excessive hip flexion) and an inability to run. Foot drop is observable. Sensory deficits include loss of sensation on the lateral side of the leg and foot. Pain and/or diminished sensation is experienced on the posterior aspect of the leg and the plantar surface of the foot (sciatica). Both the Achilles and plantar reflexes are lost. The patient is unable to stand on the toes or heels. Proprioception is lost in the foot and toes.

Short-term memory or recent memory: The ability to retrieve stored memories of events/information that occurred less than 48 hours ago. If short-term memory is impaired or lost after brain injury, it usually remains impaired.

Simultanognosia: The inability to interpret a visual stimulus as a whole.

Spasticity: A form of hypertonicity involving the inability to freely move a joint on one side of the joint—either the flexors or extensors are spastic, but not both (as in rigidity).

Staccato voice: Involves an impairment in the motor movements of speech. The patient's use of language and grammar remain intact, but the ability to clearly enunciate words is impaired. The modulation of the motor movements involved in speech is a proprioceptive or cerebellar function. Staccato voice occurs because the cerebellum cannot regulate the rate and coordination of speech motor patterns. Speech is broken, with prolonged, slow, slurred, syllables.

Stereopsis: *See* depth perception dysfunction.

Strabismus: A condition in which both eyeballs do not sit symmetrically in their orbital sockets. Strabismus may result from weakness in eyeball muscles or from an abnormality in the innervation of an eyeball muscle. If an eyeball muscle is abnormally innervated, its opposing muscle will pull the eyeball in the opposite direction. The eye deviates laterally (lateral strabismus due to a lesion of cranial nerve 3), medially (medial strabismus due to a lesion of the cranial nerve 6), or vertically and horizontally (vertical horizontal strabismus due to a lesion of the cranial nerve 4) on forward gaze.

Stocking-and-glove neuropathy: A form of neuropathy in which the distal segments of an extremity are involved first before the proximal segments. The sensation of the feet and knees are affected first, followed by the hands.

Swallowing reflex: Reflex causing involuntary swallowing when the palate is stimulated by food.

Tachykinesis: Speeded movement; commonly seen in Tourette's syndrome.

Tachyphrenia: Speeded thought; commonly seen in mania.

Tactile agnosia: The umbrella term for the inability to attach meaning to somatosensory data, or tactile data. Tactile agnosia commonly results from lesions to the secondary somatosensory area (the postcentral gyrus). One's touch and pain/temperature receptor anatomy remain intact.

Tangential speech or flight of ideas: Occurs when patients are unable to concentrate on one idea at a time for any length of time. Instead, patients jump from thought to thought, often without any obvious connection between thoughts. Their verbalizations appear to be a stream of unrelated ideas.

Tarsal tunnel syndrome: Occurs when the tibial nerve is compressed by the flexor retinaculum (the tarsal tunnel) as the nerve passes behind and beneath the medial malleolus. Damage to the tibial nerve results in a loss of plantar flexion; foot adduction and inversion; and toe flexion, abduction, and adduction. Patients are unable to stand on their toes and report that walking is both fatiguing and painful. The Achilles tendon reflex also is absent.

Tennis elbow: A severe form of pain experienced on the lateral side of the elbow commonly caused by compression of a branch of the radial nerve at the lateral epicondyle of the humerus.

Tics: Repetitive, brief, rapid, involuntary movements involving single muscles or multiple muscle groups. Tics are caused by an increased sensitivity to dopamine in the basal ganglia. With increased sensitivity to dopamine, the caudate, which normally acts like a brake on extraneous movements, cannot suppress movements like tics. A tic can involve a brief isolated movement, such as eye blinks, head jerks, or shoulder shrugs. Tics may involve a variety of sounds, such as throat clearing, grunting, or the repetition of words.

Tone: *See* muscular tone.

Topographical disorientation: Difficulty comprehending the relationship of one location to another.

Torticollis: A form of torsion dystonia characterized by contractions of the neck muscles.

Tremor: The involuntary rhythmic oscillation of joints caused by alternating contractions of opposing muscle groups. See resting tremors and intention tremors.

Truncal dystonia: A form of torsion dystonia characterized by lordosis, scoliosis, tortipelvis, and *opisthotonos* (forced flexion of the head on the chest).

Two- and three-dimensional constructional apraxia: An inability to copy or build two-dimensional and three-dimensional designs.

Two-point discrimination: The inability to determine whether one has been touched by one or two points.

Unilateral neglect: The inability to integrate and use perceptions from one side of the body and/or one side of the environment.

Vestibular-ocular reflex: A primary reflex designed to stabilize the direction of gaze compensating for head and body movement. It allows for clear vision during head and body movement.

Visual agnosia: An umbrella term for the inability to identify and recognize familiar objects and people (despite intact visual anatomical structures).

Visual attention: The ability to visually direct attention within all visual fields—superior, inferior, right, left, horizontal, and vertical.

Visual field: The space one is able to see when looking straight ahead. The normal field of vision is approximately 160° binocularly (two eyes). Monocular (one-eye) field of vision is 60° superiorly, 75° inferiorly, 60° nasally, and 100° temporally.

Visual fixation: The ability to maintain sharp focus on a moving or stationary target with both eyes precisely coordinated.

Visual pursuit: Or *tracking,* is the ability to lock onto and maintain fixation on a moving target across all visual fields.

Waiter's tip position: Results from an upper brachial plexus injury. The affected arm is held limply at the patient's side in internal rotation and adduction of the shoulder. The elbow is extended, and the forearm is pronated.

Wernicke's aphasia: Difficulty comprehending the literal interpretation of language. Wernicke's aphasia always results from a left cerebral hemisphere lesion in the brain region referred to as *Wernicke's area.*

Winging of the scapula: Occurs when the long thoracic nerve is damaged—the long thoracic nerve innervates the serratus anterior muscle. Instability of the scapula with medial rotation of the lower scapular region also commonly occur.

Wrist drop: A common sign of radial nerve damage due to denervation of the wrist extensors.

Index

Note: References in *italics* refer to figures.
References in **bold** refer to tables.